CONTENTS

List of figures	ix
Preface and acknowledgements	xi

Preamble

GENERAL INTRODUCTION 3

The background to the work 3

The scope and methods of data collection 4

The approach to data analysis 7

The style of presentation 8

Aquatic communities

INTRODUCTION TO AQUATIC COMMUNITIES 17

The sampling of aquatic vegetation 17

Data analysis and the description of aquatic communities 18

Duckweed and frogbit communities 19

Free-floating and submerged pondweed communities 20

Water-lily and pondweed vegetation with floating leaves 21

Crowfoot and starwort communities 22

Hairgrass and quillwort communities 22

Sphagnum-bladderwort vegetation. 22

KEY TO AQUATIC COMMUNITIES 23

COMMUNITY DESCRIPTIONS 27

A1 *Lemna gibba* community

    Lemnetum gibbae   Miyawaki & J.Tx. 1960 28

A2 *Lemna minor* community

    Lemnetum minoris   Soó 1947 30

A3 *Spirodela polyrhiza-Hydrocharis morsus-ranae* community 33

A4 *Hydrocharis morsus-ranae-Stratiotes aloides* community 36

Azolla filiculoides in aquatic vegetation 39

A5 *Ceratophyllum demersum* community

    Ceratophylletum demersi   Hild 1956 40

A6 *Ceratophyllum submersum* community
    Ceratophylletum submersi Den Hartog & Segal 1964   43
A7 *Nymphaea alba* community
    Nymphaeetum albae Oberdorfer & Mitarb. 1967   45
A8 *Nuphar lutea* community   48
A9 *Potamogeton natans* community   53
A10 *Polygonum amphibium* community   56
A11 *Potamogeton pectinatus-Myriophyllum spicatum* community   59
A12 *Potamogeton pectinatus* community   65
A13 *Potamogeton perfoliatus-Myriophyllum alterniflorum* community   68
A14 *Myriophyllum alterniflorum* community
    Myriophylletum alterniflori Lemée 1937   74
A15 *Elodea canadensis* community   76
Elodea nuttallii in aquatic vegetation   79
A16 *Callitriche stagnalis* community   80
A17 *Ranunculus penicillatus* ssp. *pseudofluitans* community   83
A18 *Ranunculus fluitans* community
    Ranunculetum fluitantis Allorge 1922   85
A19 *Ranunculus aquatilis* community
    Ranunculetum aquatilis Géhu 1961   87
A20 *Ranunculus peltatus* community
    Ranunculetum peltati Sauer 1947   90
A21 *Ranunculus baudotii* community
    Ranunculetum baudotii Br.-Bl. 1952   92
A22 *Littorella uniflora-Lobelia dortmanna* community   95
A23 *Isoetes lacustris/setacea* community   101
A24 *Juncus bulbosus* community   104

Swamps and tall-herb fens

INTRODUCTION TO SWAMPS AND TALL-HERB FENS   109
The sampling of swamps and tall-herb fens   109
Data analysis and the description of swamp and tall-herb fen communities   111
Swamps   115
Water-margin vegetation   117
Tall-herb fens   117

KEY TO SWAMPS AND TALL-HERB FENS   119

COMMUNITY DESCRIPTIONS   127
S1 *Carex elata* swamp
    Caricetum elatae Koch 1926   128
S2 *Cladium mariscus* swamp and sedge-beds
    Cladietum marisci Zobrist 1933 *emend.* Pfeiffer 1961   131
S3 *Carex paniculata* swamp
    Caricetum paniculatae Wangerin 1916   136
Carex appropinquata in fens   139

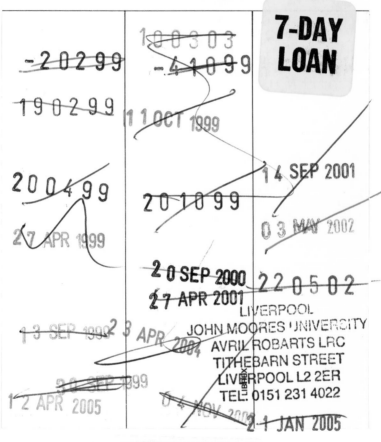

British Plant Communities

VOLUME 4

AQUATIC COMMUNITIES, SWAMPS AND TALL-HERB FENS

J. S. Rodwell (editor)
C. D. Pigott, D. A. Ratcliffe
A. J. C. Malloch, H. J. B. Birks
M. C. F. Proctor, D. W. Shimwell
J. P. Huntley, E. Radford
M. J. Wigginton, P. Wilkins

for the
U.K. Joint Nature Conservation Committee

CAMBRIDGE
UNIVERSITY PRESS

Published by the Press Syndicate of the University of Cambridge
The Pitt Building, Trumpington Street, Cambridge CB2 1RP
40 West 20th Street, New York, NY 10011-4211, USA
10 Stamford Road, Oakleigh, Melbourne 3166, Australia

First published 1995

Printed in Great Britain by Biddles Ltd, Guildford, Surrey

A catalogue record for this book is available from the British Library

Library of Congress cataloguing in publication data available

ISBN 0 521 39168 7 hardback

S4   *Phragmites australis* swamp and reed-beds
     Phragmitetum australis   (Gams 1927) Schmale 1939   140
S5   *Glyceria maxima* swamp
     Glycerietum maximae   (Nowinski 1928) Hueck 1931 *emend.* Krausch
        1965   152
S6   *Carex riparia* swamp
     Caricetum ripariae   Soó 1928   157
Carex acuta in swamps and fens   159
S7   *Carex acutiformis* swamp
     Caricetum acutiformis   Sauer 1937   160
S8   *Scirpus lacustris* ssp. *lacustris* swamp
     Scirpetum lacustris   (Allorge 1922) Chouard 1924   162
S9   *Carex rostrata* swamp
     Caricetum rostratae   Rübel 1912   166
Carex lasiocarpa in swamps and fens   170
Carex aquatilis in swamps and fens   171
S10 *Equisetum fluviatile* swamp
     Equisetetum fluviatile   Steffen 1931 *emend.* Wilczek 1935   172
S11 *Carex vesicaria* swamp
     Caricetum vesicariae   Br.-Bl. & Denis 1926   176
S12 *Typha latifolia* swamp
     Typhetum latifoliae   Soó 1927   179
S13 *Typha angustifolia* swamp
     Typhetum angustifoliae   Soó 1927   185
S14 *Sparganium erectum* swamp
     Sparganietum erecti   Roll 1938   188
S15 *Acorus calamus* swamp
     Acoretum calami   Schulz 1941   194
S16 *Sagittaria sagittifolia* swamp   197
S17 *Carex pseudocyperus* swamp   198
S18 *Carex otrubae* swamp
     Caricetum otrubae   Mirza 1978   200
S19 *Eleocharis palustris* swamp
     Eleocharitetum palustris   Schennikow 1919   203
S20 *Scirpus lacustris* ssp. *tabernaemontani* swamp
     Scirpetum tabernaemontani   Passarge 1964   206
S21 *Scirpus maritimus* swamp
     Scirpetum maritimi   (Br.-Bl. 1931) R.Tx. 1937   209
S22 *Glyceria fluitans* water-margin vegetation
     Glycerietum fluitantis   Wilczek 1935   216
S23 Other water-margin vegetation
     Glycerio-Sparganion   Br.-Bl. & Sissingh *apud* Boer 1942 *emend.* Segal   218
S24 *Phragmites australis-Peucedanum palustre* tall-herb fen
     Peucedano-Phragmitetum australis   Wheeler 1978 *emend.*   220
S25 *Phragmites australis-Eupatorium cannabinum* tall-herb fen   239
S26 *Phragmites australis-Urtica dioica* tall-herb fen   245

S27 *Carex rostrata-Potentilla palustris* tall-herb fen
     Potentillo-Caricetum rostratae   Wheeler 1980*a*   252
S28 *Phalaris arundinacea* tall-herb fen
     Phalaridetum arundinaceae   Libbert 1931   259

INDEX OF SYNONYMS TO AQUATIC COMMUNITIES,
SWAMPS AND TALL-HERB FENS   263

INDEX OF SPECIES IN AQUATIC COMMUNITIES,
SWAMPS AND TALL-HERB FENS   269

BIBLIOGRAPHY   275

FIGURES

Figure 1. Standard NVC sample card   6
Figure 2. Distribution of samples available for analysis   7
Figure 3. Floristic table for NVC community MG5 *Centaurea nigra-Cynosurus cristatus* grassland   10
Figure 4. Sampling from superimposed layers of floating and submerged aquatic vegetation among emergents at a lake edge   19
Figure 5. Key to aquatic communities   24–5
Figure 6. Zonation of aquatic, swamp and fen vegetation at Crag Lough, Northumberland   71
Figure 7. Generalised zonation across a brackish ditch   93
Figure 8. Generalised zonation of aquatic vegetation around the margins of an upland lake   98
Figure 9. Typical pattern of patchy swamp vegetation along a length of disused canal   111
Figure 10. Distribution of samples available from swamps and tall-herb fens   112

Figure 11. Pattern of coincidence between dominants and understorey assemblages in swamps and tall-herb fens   114
Figure 12. Some typical zonations with the *Phragmitetum* swamp   146
Figure 13. Mosaic of aquatic and inundation communities, swamps, fen and grassland over the draw-down zone and inlet streams of a reservoir in County Durham   173
Figure 14. Typical pattern of aquatic, swamp and fen vegetation around silting lowland pools in West Yorkshire   180
Figure 15. Variations among sequences of swamp, tall-herb fen and woodland vegetation in open-water transition mires around lowland standing and sluggish waters   231
Figure 16. Map and cross-section showing pattern of fen, fen-meadow and woodland vegetation at Market Weston Fen, Suffolk   241

PREFACE AND ACKNOWLEDGEMENTS

The appearance of Volume 4 of *British Plant Communities* brings us near the end of the publication of the results of the National Vegetation Classification. Sustaining commitment and interest through the years has been vital to the continuance of the work and, as coordinator of the NVC and editor of these volumes, I would like to record again my personal gratitude to Donald Pigott and Derek Ratcliffe, the co-chairman of our research team, to John Birks, Andrew Malloch, Michael Proctor and David Shimwell, the research supervisors and to Jacqui Huntley, Elaine Radford, Martin Wigginton and Paul Wilkins for their many and various contributions along the way. More widely, on their behalf, I wish to thank all those others who have assisted in preparing the descriptions of the aquatic, swamp and tall-herb fen vegetation that is included here.

Among the NVC team, it was David Shimwell and Elaine Radford who had particular repsonsibility for producing preliminary accounts of the swamp and fen communities and I was greatly helped by their hard work and experience. For aquatic vegetation, we also benefited from the brief secondment to the research team of Dr Michael Lock who, as well as carrying out some underwater sampling by diving, collated our own data on aquatics and helped with the early characterisation of communities.

We were fortunate, too, while working on these vegetation types to have Mrs Margaret Palmer as the Nominated Officer for the project within the Joint Nature Conservation Committee. She was already playing an important role in developing other methods for the survey of standing waters in Britain and, as well as being given access to the data which a range of projects had produced, we were much informed by her wide experience and through her comments on our work.

Among others involved in those projects, we are especially grateful to Dr Kevin Charman and Messrs Martin Alcock and Clive Doarks. Drs Terry Rowell and Nigel Holmes also made valuable comments on the manuscript at various stages and we had informative discussions, too, with Mr John Ratcliffe and Ms Ros Hattey of the then Welsh Lowland Peatland Survey, Dr John Harvey of the National Trust, the Reverend Gordon Graham of the Durham Flora Project and the late Professor David Spence, whose enthusiastic understanding of aquatic vegetation is greatly missed by all of us.

Among other contributors of data and information on the vegetation types included here, we owe a debt, as throughout our work, to Messrs Eric Birse and Jim Robertson of what was then the Macaulay Institute in Aberdeen and also, from the then West Yorkshire Biological Data Bank, to Messrs Jack Lavin and Geoffrey Wilmore. As with the mires, Dr Bryan Wheeler was unfailingly generous in allowing us access to his data on swamps and tall-herb fens from base-rich wetlands throughout Britain and always informative and constructive in his comments on the developing classification.

Within the Joint Nature Conservation Committee, Dr John Hopkins, who succeeded Margaret Palmer as Nominated Officer, has played a vital role in these closing years, helping steer the work to a successful conclusion with his commitment and understanding. At Cambridge University Press, Drs Alan Crowden and Maria Murphy have continued to provide stimulating encouragement with Mrs Jane Bulleid, as always, bringing her great skill and care to the task of sub-editing.

My appreciation of Mrs Carol Barlow, who was secretary to the NVC for more than a decade, is tempered only by my great regret at her recent departure from our team. Her constant accuracy, hard work and cheerfulness have been an enormous support throughout these critical years of the project. For this volume, I am extremely grateful to Mrs Michelle Needham for stepping into this breach and helping complete the preparation of the manuscript with care and enthusiasm.

Finally, as the work of the NVC has become more widely known through its publication in *British Plant Communities*, and its applications more diverse and productive, I am increasingly appreciative of the welcome the scheme has received from a great variety of users, within and beyond the nature conservation and countryside agencies, in this country and also abroad.

Their conviction of the value of the NVC, their enthusiastic comments and criticisms through the years, their impatience to see the project complete, these have been crucial in helping see the work through towards its end.

PREAMBLE

GENERAL INTRODUCTION

The background to the work

It is a tribute to the insight of our early ecologists that we can still return with profit to *Types of British Vegetation* which Tansley (1911) edited for the British Vegetation Committee as the first coordinated attempt to recognise and describe different kinds of plant community in this country. The contributors there wrote practically all they knew and a good deal that they guessed, as Tansley himself put it, but they were, on their own admission, far from comprehensive in their coverage. It was to provide this greater breadth, and much more detailed description of the structure and development of plant communities, that Tansley (1939) drew together the wealth of subsequent work in *The British Islands and their Vegetation*, and there must be few ecologists of the generations following who have not been inspired and challenged by the vision of this magisterial book.

Yet, partly because of its greater scope and the uneven understanding of different kinds of vegetation at the time, this is a less systematic work than *Types* in some respects: its narrative thread of explication is authoritative and engaging, but it lacks the light-handed framework of classification which made the earlier volume so very attractive, and within which the plant communities might be related one to another, and to the environmental variables which influence their composition and distribution. Indeed, for the most part, there is a rather self-conscious avoidance of the kind of rigorous taxonomy of vegetation types that had been developing for some time elsewhere in Europe, particularly under the leadership of Braun-Blanquet (1928) and Tüxen (1937). The difference in the scientific temperament of British ecologists that this reflected, their interest in how vegetation works, rather than in exactly what distinguishes plant communities from one another, though refreshing in itself, has been a lasting hindrance to the emergence in this country of any consensus as to how vegetation ought to be described, and whether it ought to be classified at all.

In fact, an impressive demonstration of the value of the traditional phytosociological approach to the description of plant communities in the British Isles was published in German after an international excursion to Ireland in 1949 (Braun-Blanquet & Tüxen 1952), but more immediately productive was a critical test of the techniques among a range of Scottish mountain vegetation by Poore (1955*a*, *b*, *c*). From this, it seemed that the really valuable element in the phytosociological method might be not so much the hierarchical definition of plant associations, as the meticulous sampling of homogeneous stands of vegetation on which this was based, and the possibility of using this to provide a multidimensional framework for the presentation and study of ecological problems. Poore & McVean's (1957) subsequent exercise in the description and mapping of communities defined using this more flexible approach then proved just a prelude to the survey of huge tracts of mountain vegetation by McVean & Ratcliffe (1962), work sponsored and published by the Nature Conservancy (as it then was) as *Plant Communities of the Scottish Highlands*. Here, for the first time, was the application of a systematised sampling technique across the vegetation cover of an extensive and varied landscape in mainland Britain, with assemblages defined in a standard fashion from full floristic data, and interpreted in relation to a complex of climatic, edaphic and biotic factors. The opportunity was taken, too, to relate the classification to other European traditions of vegetation description, particularly that developed in Scandinavia (Nordhagen 1943, Dahl 1956).

McVean & Ratcliffe's study was to prove a continual stimulus to the academic investigation of our mountain vegetation and of abiding value to the development of conservation policy, but their methods were not extended to other parts of the country in any ambitious sponsored surveys in the years immediately following. Despite renewed attempts to commend traditional phytosociology, too (Moore 1962), the attraction of this whole approach was overwhelmed for many by the heated debates that preoccupied British plant ecologists in the 1960s, on the issues of objectivity in the sampling and sorting of data, and the respective values of classification or ordination as analytical techniques. Others, though, found it perfectly possible to integrate multivariate analysis into phytosociological survey, and demonstrated the advantage of computers for the display and interpretation of ecological data, rather than the simple testing of methodologies (Ivimey-Cook & Proctor 1966). New generations of research students also began to draw inspiration from the Scottish and Irish initiatives by applying phytosociology to the solving of particular descriptive and interpretive problems, such as variation among British calcicolous grasslands (Shimwell 1968a), heaths (Bridgewater 1970), rich fens (Wheeler 1975) and salt-marshes (Adam 1976), the vegetation history of Skye (Birks 1969), Cornish cliffs (Malloch 1970) and Upper Teesdale (Bradshaw & Jones 1976). Meanwhile, too, workers at the Macaulay Institute in Aberdeen had been extending the survey of Scottish vegetation to the lowlands and the Southern Uplands (Birse & Robertson 1976, Birse 1980, 1984).

With an accumulating volume of such data and the appearance of uncoordinated phytosociological perspectives on different kinds of British vegetation, the need for an overall framework of classification became ever more pressing. For some, it was also an increasingly urgent concern that it still proved impossible to integrate a wide variety of ecological research on plants within a generally accepted understanding of their vegetational context in this country. Dr Derek Ratcliffe, as Scientific Assessor of the Nature Conservancy's Reserves Review from the end of 1966, had encountered the problem of the lack of any comprehensive classification of British vegetation types on which to base a systematic selection of habitats for conservation. This same limitation was recognised by Professor Sir Harry Godwin, Professor Donald Pigott and Dr John Phillipson who, as members of the Nature Conservancy, had been asked to read and comment on the Reserves Review. The published version, A Nature Conservation Review (Ratcliffe 1977), was able to base the description of only the lowland and upland grasslands and heaths on a phytosociological treatment. In 1971, Dr Ratcliffe, then Deputy Director (Scientific) of the Nature Conservancy, in proposals for development of its research programme, drew attention to 'the need for a national and systematic phytosociological treatment of British vegetation, using standard methods in the field and analysis/classification of the data'. The intention of setting up a group to examine the issue lapsed through the splitting of the Conservancy which was announced by the Government in 1972. Meanwhile after discussions with Dr Ratcliffe, Professor Donald Pigott of the University of Lancaster proposed to the Nature Conservancy a programme of research to provide a systematic and comprehensive classification of British plant communities. The new Nature Conservancy Council included it as a priority item within its proposed commissioned research programme. At its meeting on 24 March 1974, the Council of the British Ecological Society welcomed the proposal. Professor Pigott and Dr Andrew Malloch submitted specific plans for the project and a contract was awarded to Lancaster University, with sub-contractual arrangements with the Universities of Cambridge, Exeter and Manchester, with whom it was intended to share the early stages of the work. A coordinating panel was set up, jointly chaired by Professor Pigott and Dr Ratcliffe, and with research supervisors from the academic staff of the four universities, Drs John Birks, Michael Proctor and David Shimwell joining Dr Malloch. At a later stage, Dr Tim Bines replaced Dr Ratcliffe as nominated officer for the NCC, and Miss Lynne Farrell succeeded him in 1985.

With the appointment of Dr John Rodwell as full-time coordinator of the project, based at Lancaster, the National Vegetation Classification began its work officially in August 1975. Shortly afterwards, four full-time research assistants took up their posts, one based at each of the universities: Mr Martin Wigginton, Miss Jacqueline Paice (later Huntley), Mr Paul Wilkins and Dr Elaine Grindey (later Radford). These remained with the project until the close of the first stage of the work in 1980, sharing with the coordinator the tasks of data collection and analysis in different regions of the country, and beginning to prepare preliminary accounts of the major vegetation types. Drs Michael Lock and Hilary Birks and Miss Katherine Hearn were also able to join the research team for short periods of time. After the departure of the research assistants, the supervisors supplied Dr Rodwell with material for writing the final accounts of the plant communities and their integration within an overall framework. With the completion of this charge in 1989, the handover of the manuscript for publication by the Cambridge University Press began.

The scope and methods of data collection

The contract brief required the production of a classification with standardised descriptions of named and systematically arranged vegetation types and, from the beginning, this was conceived as something much more than an annotated list of interesting and unusual plant communities. It was to be comprehensive in its coverage, taking in the whole of Great Britain but not Northern Ireland, and including vegetation from all natural, semi-natural and major artificial habitats. Around the maritime fringe, interest was to extend up to the start of the truly marine zone, and from there to the tops of our remotest mountains, covering virtually all terrestrial plant communities and those of brackish and fresh waters, except where non-vascular plants were the dominants. Only short-term leys were specifically excluded, and, though care was to be taken to sample more pristine and long-established kinds of vegetation, no undue attention was to be given to assemblages of rare plants or to especially rich and varied sites. Thus widespread and dull communities from improved pastures, plantations, run-down mires and neglected heaths were to be extensively sampled, together with the vegetation of paths, verges and recreational swards, walls, man-made waterways and industrial and urban wasteland.

For some vegetation types, we hoped that we might be able to make use, from early on, of existing studies, where these had produced data compatible in style and quality with the requirements of the project. The contract envisaged the abstraction and collation of such material from both published and unpublished sources, and discussions with other workers involved in vegetation survey, so that we could ascertain the precise extent and character of existing coverage and plan our own sampling accordingly. Systematic searches of the literature and research reports revealed many data that we could use in some way and, with scarcely a single exception, the originators of such material allowed us unhindered access to it. Apart from the very few classic phytosociological accounts, the most important sources proved to be postgraduate theses, some of which had already amassed very comprehensive sets of samples of certain kinds of vegetation or from particular areas, and these we were generously permitted to incorporate directly.

Then, from the NCC and some other government agencies, or from individuals who had been engaged in earlier contracts for them, there were some generally smaller bodies of data, occasionally from reports of extensive surveys, more usually from investigations of localised areas. Published papers on particular localities, vegetation types or individual species also provided small numbers of samples. In addition to these sources, the project was able to benefit from and influence ongoing studies by institutions and individuals, and itself to stimulate new work with a similar kind of approach among university researchers, NCC surveyors, local flora recorders and a few suitably qualified amateurs. An initial assessment and annual monitoring of floristic and geographical coverage were designed to ensure that the accumulating data were fairly evenly spread, fully representative of the range of British vegetation, and of a consistently high quality. Full details of the sources of the material, and our acknowledgements of help, are given in the preface and introduction to each volume.

Our own approach to data collection was simple and pragmatic, and a brief period of training at the outset ensured standardisation among the team of five staff who were to carry out the bulk of the sampling for the project in the field seasons of the first four years, 1976–9.

The thrust of the approach was phytosociological in its emphasis on the systematic recording of floristic information from stands of vegetation, though these were chosen solely on the basis of their relative homogeneity in composition and structure. Such selection took a little practice, but it was not nearly so difficult as some critics of this approach imply, even in complex vegetation, and not at all mysterious. Thus, crucial guidelines were to avoid obvious vegetation boundaries or unrepresentative floristic or physiognomic features. No prior judgements were necessary about the identity of the vegetation type, nor were stands ever selected because of the presence of species thought characteristic for one reason or another, nor by virtue of any observed uniformity of the environmental context.

From within such homogeneous stands of vegetation, the data were recorded in quadrats, generally square unless the peculiar shape of stands dictated otherwise. A relatively small number of possible sample sizes was used, determined not by any calculation of minimal areas, but by the experienced assessment of their appropriateness to the range of structural scale found among our plant communities. Thus plots of 2×2 m were used for most short, herbaceous vegetation and dwarf-shrub heaths, 4×4 m for taller or more open herb communities, sub-shrub heaths and low woodland field layers, 10×10 m for species-poor or very tall herbaceous vegetation or woodland field layers and dense scrub, and 50×50 m for sparse scrub, and woodland canopy and understorey. Linear vegetation, like that in streams and ditches, on walls or from hedgerow field layers, was sampled in 10 m strips, with 30 m strips for hedgerow shrubs and trees. Quadrats of 1×1 m were rejected as being generally inadequate for representative sampling, although some bodies of existing data were used where this, or other sizes different from our own, had been employed. Stands smaller than the relevant sample size were recorded in their entirety, and mosaics were treated as a single vegetation type where they were repeatedly encountered in the same form, or where their scale made it quite impossible to sample their elements separately.

Samples from all different kinds of vegetation were recorded on identical sheets (Figure 1). Priority was always given to the accurate scoring of all vascular plants, bryophytes and macrolichens (*sensu* Dahl 1968), a task which often required assiduous searching in dense and complex vegetation, and the determination of difficult plants in the laboratory or with the help of referees. Critical taxa were treated in as much detail as possible though, with the urgency of sampling, certain groups, like the brambles, hawkweeds, eyebrights and dandelions, often defeated us, and some awkward bryophytes and crusts of lichen squamules had to be referred to just a genus. It is more than likely, too, that some very diminutive mosses and especially hepatics escaped notice in the field and, with much sampling taking place in summer, winter annuals and vernal perennials might have been missed on occasion. In general, nomenclature for vascular plants follows *Flora Europaea* (Tutin *et al.* 1964 *et seq.*) with Corley & Hill (1981) providing the authority for bryophytes and Dahl (1968) for lichens. Any exceptions to this, and details of any difficulties with sampling or identifying particular plants, are given in the introductions to each of the major vegetation types.

A quantitative measure of the abundance of every taxon was recorded using the Domin scale (*sensu* Dahl & Hadač 1941), cover being assessed by eye as a vertical projection on to the ground of all the live, above-ground parts of the plants in the quadrat. On this scale:

Cover of 91–100%	is recorded as Domin	10
76–90%		9
51–75%		8
34–50%		7
26–33%		6
11–25%		5
4–10%		4
<4%	with many individuals	3
	with several individuals	2
	with few individuals	1

Figure 1. Standard NVC sample card.

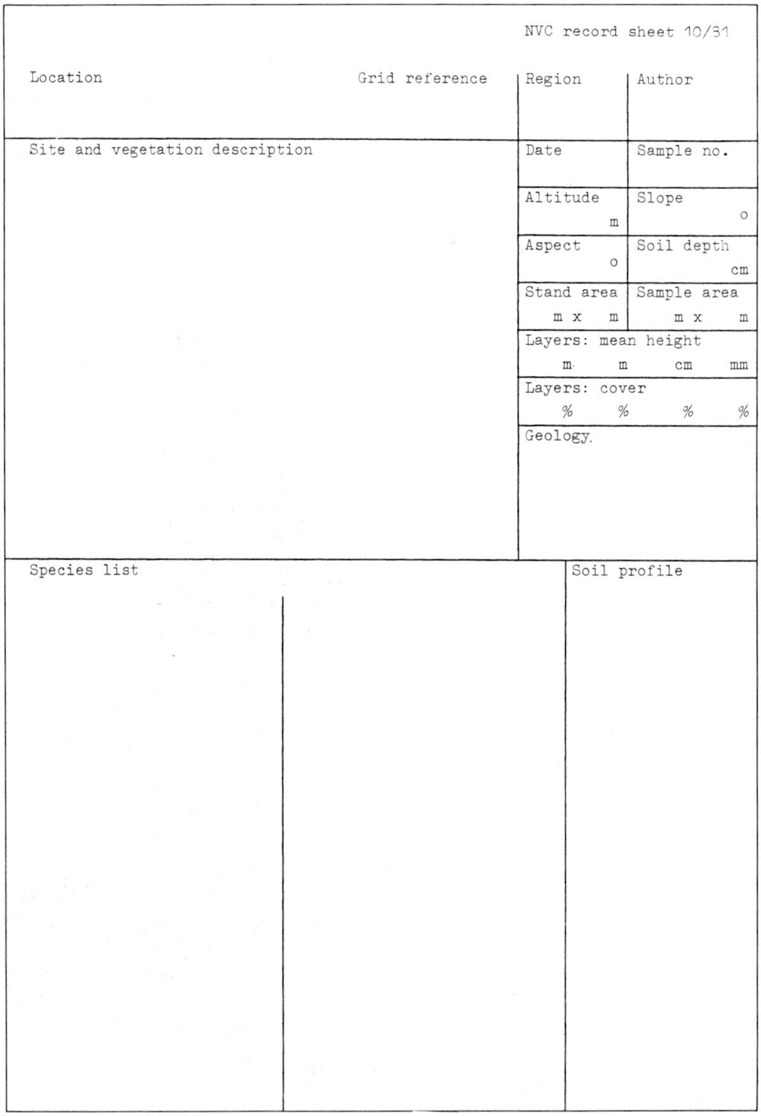
NVC record sheet 10/81

Location
Grid reference
Region
Author
Site and vegetation description
Date
Sample no.
Altitude m
Slope o
Aspect o
Soil depth cm
Stand area m x m
Sample area m x m
Layers: mean height m m cm mm
Layers: cover % % % %
Geology
Species list
Soil profile

entials, as with *Festuca pratensis* here: it is not often found in *Centaurea-Cynosurus* grassland but, when it does occur, it is generally in this first sub-type.

The species group *Galium verum* to *Festuca ovina* helps to distinguish the second sub-community from the first, though again there is some variation in the strength of association between these preferentials and the vegetation type, with *Achillea millefolium* being less markedly diagnostic than *Trisetum flavescens* and, particularly, *G. verum*. There are also important negative features, too, because, although some plants typical of the first and third sub-communities, such as *Lolium* and *Prunella vulgaris*, remain quite common here, the disappearance of others, like *Lathyrus pratensis*, *Danthonia decumbens*, *Potentilla erecta* and *Succisa pratensis* is strongly diagnostic. Similarly, with the third sub-community, there is that same mixture of positive and negative characteristics, and there is, among all the groups of preferentials, that same variation in abundance as is found among the constants and companions. Thus, some plants which can be very marked preferentials, are always of rather low cover, as with *Prunella*, whereas others, like *Agrostis stolonifera*, though diagnostic at low frequency, can be locally plentiful.

For the naming of the sub-communities, we have generally used the most strongly preferential species, not necessarily those most frequent in the vegetation type. Sometimes, sub-communities are characterised by no floristic features over and above those of the community as a whole, in which case there will be no block of preferentials on the table. Usually, such vegetation types have been called Typical, although we have tried to avoid this epithet where the sub-community has a very restricted or eccentric distribution.

The tables organise and summarise the floristic variation which we encountered in the vegetation sampled: the text of the community accounts attempts to expound and interpret it in a standardised descriptive format. For each community, there is first a synonymy section which lists those names applied to that particular kind of vegetation where it has figured in some form or another in previous surveys, together with the name of the author and the date of ascription. The list is arranged chronologically, and it includes references to important unpublished studies and to accounts of Irish and Continental associations where these are obviously very similar. It is important to realise that very many synonyms are inexact, our communities corresponding to just part of a previously described vegetation type, in which case the initials *p.p.* (for *pro parte*) follow the name, or being subsumed within an older, more broadly defined unit. Despite this complexity, however, we hope that this section, together with that on the affinities of the vegetation (see below), will help readers translate our scheme into terms with which they may have been long familiar.

A special attempt has been made to indicate correspondence with popular existing schemes and to make sense of venerable but ill-defined terms like 'herb-rich meadow', 'oakwood' or 'general salt-marsh'.

There then follow a list of the constant species of the community, and a list of the rare vascular plants, bryophytes and lichens which have been encountered in the particular vegetation type, or which are reliably known to occur in it. In this context, 'rare' means, for vascular plants, an A rating in the *Atlas of the British Flora* (Perring & Walters 1962), where scarcity is measured by occurrence in vice-counties, or inclusion on lists compiled by the NCC of plants found in less than one hundred 10×10 km squares. For bryophytes, recorded presence in under 20 vice-counties has been used as a criterion (Corley & Hill 1981), with a necessarily more subjective estimate for lichens.

The first substantial section of text in each community description is an account of the physiognomy, which attempts to communicate the feel of the vegetation in a way which a tabulation of data can never do. Thus, the patterns of frequency and abundance of the different species which characterise the community are here filled out by details of the appearance and structure, variation in dominance and the growth form of the prominent elements of the vegetation, the physiognomic contribution of subordinate plants, and how all these components relate to one another. There is information, too, on important phenological changes that can affect the vegetation through the seasons and an indication of the structural and floristic implications of the progress of the life cycle of the dominants, any patterns of regeneration within the community or obvious signs of competitive interaction between plants. Much of this material is based on observations made during sampling, but it has often been possible to incorporate insights from previous studies, sometimes as brief interpretive notes, in other cases as extended treatments of, say, the biology of particular species such as *Phragmites australis* or *Ammophila arenaria*, the phenology of winter annuals or the demography of turf perennials. We trust that this will help demonstrate the value of this kind of descriptive classification as a framework for integrating all manner of autecological studies (Pigott 1984).

Some indication of the range of floristic and structural variation within each community is given in the discussion of general physiognomy, but where distinct sub-communities have been recognised these are each given a descriptive section of their own. The sub-community name is followed by any synonyms from previous studies, and by a text which concentrates on pointing up the particular features of composition and organisation which distinguish it from the other sub-communities.

Passing reference is often made in these portions of the community accounts to the ways in which the nature of the vegetation reflects the influence of environmental factors upon it, but extended treatment of this is reserved for a section devoted to the habitat. An opening paragraph here attempts to summarise the typical conditions which favour the development and maintenance of the vegetation type, and the major factors which control floristic and structural variation within it. This is followed by as much detail as we have at the present time about the impact of particular climatic, edaphic and biotic variables on the community, or as we suppose to be important to its essential character and distribution. With climate, for example, reference is very frequently made to the influence on the vegetation of the amount and disposition of rainfall through the year, the variation in temperature season by season, differences in cloud cover and sunshine, and how these factors interact in the maintenance of regimes of humidity, drought or frosts. Then, there can be notes of effects attributable to the extent and duration of snow-lie or to the direction and strength of winds, especially where these are icy or salt-laden. In each of these cases, we have tried to draw upon reputable sources of data for interpretation, and to be fully sensitive to the complex operation of topographic climates, where features like aspect and altitude can be of great importance, and of regional patterns, where concepts like continental, oceanic, montane and maritime climates can be of enormous help in understanding vegetation patterns.

Commonly, too, there are interactions between climate and geology that are best perceived in terms of variations in soils. Here again, we have tried to give full weight to the impact of the character of the landscape and its rocks and superficials, their lithology and the ways in which they weather and erode in the processes of pedogenesis. As far as possible, we have employed standardised terminology in the description of soils, trying at least to distinguish the major profile types with which each community is associated, and to draw attention to the influence of its floristics and structure of processes like leaching and podzolisation, gleying and waterlogging, parching, freeze-thaw and solifluction, and inundation by fresh- or salt-waters.

With very many of the communities we have distinguished, it is combinations of climatic and edaphic factors that determine the general character and possible range of the vegetation, but we have often also been able to discern biotic influences, such as the effects of wild herbivores or agents of dispersal, and there are very few instances where the impact of man cannot be seen in the present composition and distribution of the plant communities. Thus, there is frequent reference to the role which treatments such as grazing, mowing and burning have on the floristics and physiognomy of the vegetation, to the influence of manuring and other kinds of eutrophication, of draining and re-seeding for agri-culture, of the cropping and planting of trees, of trampling or other disturbance, and of various kinds of recreation.

The amount and quality of the environmental information on which we have been able to draw for interpreting such effects has been very variable. Our own sampling provided just a spare outline of the physical and edaphic conditions at each location, data which we have summarised where appropriate at the foot of the floristic tables; existing sources of samples sometimes offered next to nothing, in other cases very full soil analyses or precise specifications of treatments. In general, we have used what we had, at the risk of great unevenness of understanding, but have tried to bring some shape to the accounts by dealing with the environmental variables in what seems to be their order of importance, irrespective of the amount of detail available, and by pointing up what can already be identified as environmental threats. We have also benefited by being able to draw on the substantial literature on the physiology and reproductive biology of individual species, on the taxonomy and demography of plants, on vegetation history and on farming and forestry techniques. Sometimes, this information provides little more than a provisional substantiation of what must remain for the moment an interpretive hunch. In other cases, it has enabled us to incorporate what amount to small essays on, for example, the past and present role of *Tilia cordata* in our woodlands with variation in climate, the diverse effects of dunging by rabbits, sheep and cattle on calcicolous swards, or the impact of burning on *Calluna-Arctostaphylos* heath on different soils in a boreal climate. Debts of this kind are always acknowledged in the text and, for our part, we hope that the accounts indicate the benefits of being able to locate experimental and historical studies on vegetation within the context of an understanding of plant communities (Pigott 1982).

Mention is often made in the discussion of the habitat of the ways in which stands of communities can show signs of variation in relation to spatial environmental differences, or the beginnings of a response to temporal changes in conditions. Fuller discussion of zonations to other vegetation types follows, with a detailed indication of how shifts in soil, microclimate or treatment affect the composition and structure of each community, and descriptions of the commonest patterns and particularly distinctive ecotones, mosaics and site types in which it and any sub-communities are found. It has also often been possible to give some fuller and more ordered account of the ways in which vegetation types can change through time, with invasion of newly available ground, the progression of communities to maturity, and their regeneration and replacement. Some attempt has been made to identify climax vegetation types and major lines of succession, but we have always been wary of the temptation to extrapolate from spatial patterns to temporal sequences. Once more, we have tried to incorporate the results of existing observational and experimental studies, including some of the classic accounts of patterns and processes among British vegetation, and to point up the great advantages of a reliable scheme of classification as a basis for the monitoring and management of plant communities (Pigott 1977).

Throughout the accounts, we have referred to particular sites and regions wherever we could, many of these visited and sampled by the team, some the location of previous surveys, the results of which we have now been able to redescribe in the terms of the classification we have erected. In this way, we hope that we have begun to make real a scheme which might otherwise remain abstract. We have also tried in the habitat section to provide some indications of how the overall ranges of the vegetation types are determined by environmental conditions. A separate paragraph on distribution summarises what we know of the ranges of the communities and sub-communities, then maps show the location, on the 10×10 km national grid, of the samples that are available to us for each. Much ground, of course, has been thinly covered, and sometimes a dense clustering of samples can reflect intensive sampling rather than locally high frequency of a vegetation type. However, we believe that all the maps we have included are accurate in their general indication of distributions, and we hope that this exercise might encourage the production of a comprehensive atlas of British plant communities.

The last section of each community description considers the floristic affinities of the vegetation types in the scheme, and expands on any particular problems of synonymy with previously described assemblages. Here, too, reference is often given to the equivalent or most closely related association in Continental phytosociological classifications and an attempt made to locate each community in an existing alliance. Where the fuller account of British vegetation that we have been able to provide necessitates a revision of the perspective on European plant communities as a whole, some suggestions are made as to how this might be achieved.

Meanwhile, each reader will bring his or her own needs and commitment to this scheme and perhaps be dismayed by its sheer size and apparent complexity. For those requiring some guidance as to the scope of each volume and the shape of that part of the classification with which it deals, the introductions to the major vegetation types will provide an outline of the variation and how it has been treated. The contents page will then give directions to the particular communities of interest. For readers less sure of the identity of the vegetation types with which they are dealing, a key is provided to each major group of communities which should enable a set of similar samples organised into a constancy table to be taken through a series of questions to a reasonably secure diagnosis. The keys, though, are not infallible short cuts to identification and must be used in conjunction with the floristic tables and community descriptions. An alternative entry to the scheme is provided by the species index which lists the occurrences of all taxa in the communities in which we have recorded them. There is also an index of synonyms which should help readers find the equivalents in our classification of vegetation types already familiar to them.

Finally, we hope that whatever the needs, commitments or even prejudices of those who open these volumes, there will be something here to inform and challenge everyone with an interest in vegetation. We never thought of this work as providing the last word on the classification of British plant communities: indeed, with the limited resources at our disposal, we knew it could offer little more than a first approximation. However, we do feel able to commend the scheme as essentially reliable. We hope that the broad outlines will find wide acceptance and stand the test of time, and that our approach will contribute to setting new standards of vegetation description. At the same time, we have tried to be honest about admitting deficiencies of coverage and recognising much unexplained floristic variation, attempting to make the accounts sufficiently open-textured that new data might be readily incorporated and ecological puzzles clearly seen and pursued. For the classification is meant to be not a static edifice, but a working tool for the description, assessment and study of vegetation. We hope that we have acquitted ourselves of the responsibilities of the contract brief and the expectations of all those who have encouraged us in the task, such that the work might be thought worthy of standing in the tradition of British ecology. Most of all, we trust that our efforts do justice to the vegetation which, for its own sake, deserves understanding and care.

have characterised are equivalent to the 'societies' recognised by authors like Spence (1964) and Westhoff & den Held (1969). In other cases, vegetation of this kind has found a place as a species-poor sub-community within a broader assemblage defined here. Also problematic to classify are mixtures or intimate mosaics of such dominants, again with few if any associates. Here, there is little alternative to grouping samples according to the proportions of the dominant species.

The communities can be considered under six general headings: surface and sub-surface duckweed and frogbit vegetation of moderately-rich to eutrophic standing waters (4 communities), free-floating or rooted and submerged pondweed vegetation (7 communities), rooted water-lily and pondweed vegetation with floating leaves (6 communities), crowfoot and starwort vegetation of running waters (3 communities), submerged

swards of quillworts and hairgrass (2 communities) and free-floating vegetation of impoverished base-poor standing waters (1 community). All these communities are described in the usual style, although maps of their occurrence have been omitted because the often patchy cover of our sampling could give a misleading impression of distribution patterns. We have also included notes on the occurrence of *Azolla filiculoides* and *Elodea nuttallii* among the aquatic assemblages characterised.

Duckweed and frogbit communities

Shallow standing or sluggish waters richer in nutrients, in the sheltered parts of big lakes but more commonly in small lowland pools and ponds, ditches and quiet canals, characteristically develop floating carpets of duckweeds. Most widespread is the *Lemna minor* community (A2 *Lemnetum minoris* Soó 1947) with the *Lemna gibba* community (A1 *Lemnetum gibbae* Miyawaki & J.Tx. 1960) more strictly confined to the warmer south and east of Britain, somewhat more base- and nutrient-

Figure 4. Sampling from superimposed layers of floating and submerged aquatic vegetation among emergents at a lake edge.

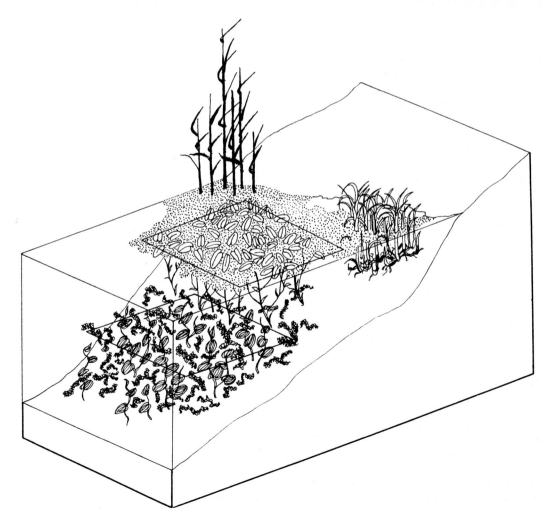

demanding and more tolerant of recently disturbed and unstable conditions.

Lemna trisulca frequently occurs floating beneath these other duckweeds, particularly with *L. minor*, and it can be accompanied by the thalloid liverworts *Riccia fluitans* and *Ricciocarpus natans*. Further sampling in very sheltered waters and their muddy margins may enable a *Riccietum fluitantis* to be recognised as, for example, in Westhoff & den Held (1969) but, for the moment, we have retained such vegetation in the *Lemnetum minoris*.

Spirodela polyrhiza, too, is sometimes a feature of these simpler mats, but richer assemblages of all four common duckweed species, occasionally also with the rare *Wolffia arrhiza*, have been placed in a separate *Spirodela-Hydrocharis morsus-ranae* community (A3). Along with *Hydrocharis*, there are quite often associated submerged masses of *Elodea canadensis* and *Ceratophyllum demersum* together with stocky *Berula erecta*. This community is confined to unpolluted clear and unshaded lowland waters, sometimes quite calcareous or even slightly saline and is becoming increasingly local among dyke systems, ponds and quiet canals.

Hydrocharis, *Lemna minor* and *L. trisulca* all remain constant, along with suspended masses of *Elodea* and *Ceratophyllum demersum*, in a further distinctive and very local assemblage, the *Hydrocharis-Stratiotes aloides* community (A4). As well as the striking rarity *Stratiotes*, this vegetation also often has floating *Utricularia vulgaris* and *Callitriche platycarpa* as well as some rooted aquatics like *Nuphar lutea* and *Myriophyllum verticillatum*. Probably at one time more widespread in clear, base-rich waters through the warmer eastern lowlands of England, this community is now restricted to just a few dyke systems, mostly in Broadland. It seems unlikely, though, that its decline is due simply to nutrient enrichment of these waters as it tends to favour distinctly mesotrophic systems. There is some disagreement as to whether it should be placed among the Lemnetalia or in a separate alliance, the Hydrocharition, among the Potametea.

Another distinctive floating aquatic in eutrophic ponds, ornamental pools and stagnant dykes in lowland Britain is the tropical fern *Azolla filiculoides*, widely naturalised over the last century and prodigiously expansive when conditions are favourable. Again, we have not recognised a separate assemblage but have provided a note on its habitat and affinities.

All of these vegetation types are characteristic of the NCC 'southern' eutrophic Standing Water Site types 8–10 (Palmer *et al.* 1992).

Free-floating and submerged pondweed communities

Submerged aquatics, free-floating or rooted, sometimes figure in inextricable mixtures with richer duckweed assemblages but they can occur, too, in separate communities, colonising standing and sluggish waters that may be of considerable depth.

Two of the commonest and most widespread plants of importance among this group are *Ceratophyllum demersum* and *Elodea canadensis* which can occur intermixed one with the other in all proportions and also in purer stands of a *Ceratophyllum demersi* Hild 1956 (A5) or an *Elodea canadensis* community (A15). Both these kinds of vegetation are characteristic of more eutrophic waters in ponds and lakes, dykes, canals and slow-moving streams and rivers, particularly in the warmer lowlands of Britain, with the *Ceratophylletum* having the added advantage of being shade-tolerant.

E. canadensis, of course, is an introduction, first reliably recorded in Britain only in 1842, since when it has spread widely, both nationally and on a local scale with a characteristic pattern of rapid luxuriance and subsequent decline. Its occurrence is now also complicated by the realisation that *E. nuttallii*, probably a much more recent introduction and morphologically varied enough to confuse the unwary, is widespread too. It can occur with *E. canadensis*, though both are rarely abundant together, and also in the *Ceratophylletum* and other pondweed vegetation, but we have insufficient data to characterise a distinct community.

The softer and fresher green relative of *Ceratophyllum demersum*, *C. submersum*, is also sometimes found as dense masses of unanchored shoots, but is generally found with *Potamogeton pectinatus* and the rare *Ranunculus baudotii*, together with occasional *Myriophyllum spicatum* and *Zannichellia palustris* in a distinctive *Ceratophyllum submersum* community (A6 *Ceratophyllum submersi* Den Hartog & Segal 1964). This vegetation is characteristic of standing or sluggish, eutrophic and usually brackish waters in pools and dykes of coastal and estuarine marshes around the sub-maritime fringe of Sussex, Kent and Essex.

In similar habitats, but a little more widespread around the coast of southern Britain, is found the *Ranunculus baudotii* community (A21 *Ranunculetum baudotii* Br.-Bl. 1952) where *C. submersum*, *P. pectinatus* and *Z. palustris* remain as frequent associates, but where dominance shifts to *R. baudotii*. This vegetation is also able to persist fragmentarily on the moist margins of fluctuating waters disturbed by grazing stock and wildfowl.

In contrast to the other communities in this group, the *Ranunculetum baudotii* is generally placed among the tassleweed vegetation of the Ruppion maritimae. The *Ceratophylletum demersi*, *Ceratophylletum submersi* and *Elodea canadensis* community, together with the four assemblages described below, have traditionally been grouped among the pondweed assemblages of the Parvopotamion, recently subsumed in a broader Potamion, within the Potametea.

Some of these remaining communities are very simple in composition and they overlap in their habitat requirements with the above. The *Potamogeton pectinatus* community (A12), for example, can have some *Myriophyllum spicatum*, *Ceratophyllum demersum* and *Zannichellia palustris*, but is often species-poor vegetation strongly dominated by bushy clumps of *P. pectinatus*. It extends into brackish habitats but is widespread through the warmer lowlands of Britain in all manner of eutrophic waters, standing, sluggish and quite fast-moving and, being tolerant of both enrichment and pollution, has become increasingly common in pools and dykes in agricultural and industrial landscapes.

In more base-rich and cleaner waters, this assemblage grades to the *Potamogeton pectinatus-Myriophyllum spicatum* community (A11) which includes much of the species-rich pondweed vegetation for which we had samples. This is a very broadly-defined group which perhaps subsumes more than one assemblage but, as characterised here, the frequent associates include *Potamogeton pusillus*, *P. crispus* and *P. natans*, with *P. lucens*, *P. berchtoldii*, *P. perfoliatus* and *P. obtusifolius* occasional in more diverse stands, *P. filiformis*, *P. friesii* and *P. gramineus* more restricted in occurrence. Although tolerant of modest eutrophication, when *Elodea* spp. tend to become more abundant, this vegetation has become increasingly impoverished with pollution and turbidity of lowland waters. It occurs widely but locally now in dykes, canals, ponds and lakes, and in streams and rivers that are not too swift or spatey. Some of these habitats fall within the mesotrophic NCC Standing Water Site type 5, but the community also includes a very distinctive kind of pondweed vegetation characteristic of the marly waters of the machair lochs of the western seaboard of Scotland grouped in Site type 7 (Palmer *et al.* 1992).

Typical of more base-poor, shallow to quite deep waters of only moderate trophic state is the *Potamogeton perfoliatus-Myriophyllum alterniflorum* community (A13). The broader-leaved *P. perfoliatus* and *P. gramineus* dominate among the pondweeds here with *P. berchtoldii*, *P. obtusifolius*, *P. pusillus*, *P. natans* and *P. filiformis* more occasional. *Myriophyllum alterniflorum* is the characteristic milfoil and *Littorella uniflora* is a constant, sometimes thickening up to a dense sward. Suitable conditions for this vegetation type are met only locally in the more fertile lowlands of Britain but it occurs widely through the north and west, in pools and lakes, and the middle to lower reaches of rivers where the accumulation of fine to moderately coarse sediments brings modest nutrient- and base-enrichment. Such situations are represented among the mesotrophic and oligotrophic NCC Standing Water Site types 2, 3 and 5, though more peaty substrates in machair lochs (NCC type 4: Palmer *et al.* 1992) can also support this kind of pondweed vegetation closely juxtaposed with its more basiphilous counterpart. In fact, *P. pusillus* and *P. filiformis* may be part of a separate distinct assemblage characteristic of such conditions.

Where conditions are less fertile and somewhat more exposed to wave action in lakes or swift and spatey waters in rivers, the broader-leaved pondweeds become battered and torn and the cover thins to mixtures of *M. alterniflorum* and *Littorella*. In the *Myriophyllum alterniflorum* community (A14 *Myriophylletum alterniflori* Lemée 1937), which can persist even in very stony situations, the milfoil tends to dominate but other occasional associates include *Lobelia dortmanna*, *Callitriche hamulata* and *Juncus bulbosus*, bringing the composition close to the Littorelletea swards.

Water-lily and pondweed vegetation with floating leaves

From shallower standing and moving waters or deeper lakes where sediments have accumulated, we have characterised a variety of communities of rooted aquatics with floating leaves. These vegetation types have generally been grouped in the Nymphaeion and, producing often crowded masses of densely-shading foliage, they are often poor in associated species, sometimes little more than monodominant stands.

Most tolerant of nutrient-poor conditions and of turbulent waters is the *Potamogeton natans* community (A9) which, where it extends into sandy and gravelly lakes or spatey streams in north-west Britain, can have a sparse understorey of the species of the *Myriophylletum*. In shallower and more sluggish, though not necessarily richer, waters and extending on to the periodically flooded surrounds and banks of ponds, reservoirs and more mature rivers, is the amphibious *Polygonum amphibium* community (A10).

With their much bulkier growth, the water-lilies tend to favour deeper waters. The *Nymphaea alba* community (A7 *Nymphaeetum albae* Oberdorfer & Mitarb. 1967) is typical of more oligotrophic and base-poor conditions, most strikingly developed in acidic and often peaty waters in the sheltered bays of upland lakes (NCC Standing Water Site types 2 and 3). Here it may grow alone and vigorous or, in more dystrophic shallows and soakways, with *Potamogeton polygonifolius*, *P. natans*, *Juncus bulbosus* and *Myriophyllum alterniflorum*.

N. alba has been widely planted and it can also be found in more eutrophic ponds, lakes and canals through the lowland south and east. More characteristic there, though (among NCC Standing Water Site type 5), is the *Nuphar lutea* community (A8) which, with the extensive rhizomes and long petioles of the dominant can occur in up to 3 m of water at the outer limit of floating-leaved vegetation. Many stands are very poor in associates, but some have a mixture of other water lilies, *Nymphaea alba* but also *Nuphar pumila*, the hybrid *N. × spennerana* and *Nymphoides peltata*. Yet others, extending the range of the community into richer, unpolluted and often marly waters, have an abundant understorey with *Callitriche stagnalis* and *Zannichellia palustris*.

Also usually included in the Nymphaeion are crowfoot communities of the shallows and surrounds of mesotrophic to eutrophic waters. More tolerant of turbulence and extending therefore into quite fast-moving streams in the upland fringes is the *Ranunculus aquatilis* community (A19 *Ranunculetum aquatilis* Géhu 1961) but the ability of this plant to survive periodic or seasonal drying also gives the community a place around fluctuating pond margins and the upper reaches of chalk streams.

The *Ranunculus peltatus* community (A20 *Ranunculetum peltati*   Sauer 1947) is more consistently associated with sluggish or still waters, again sometimes fluctuating, and it seems to tolerate more calcareous and impoverished situations than the *Ranunculetum aquatilis*. Both communities may benefit from the disturbance which stock provide around streams and ponds.

Crowfoot and starwort communities

The distinctive submerged festoons of water-crowfoots growing in moderately fast to quite swift streams and rivers have usually been placed in a third alliance of the Potametea, the Ranunculion fluitantis.

The *Ranunculus fluitans* community itself (A18 *Ranunculetum fluitantis*   Allorge 1922) is a striking sight with its bushy trails of the dominant growing up to 6 m long in usually only moderately fertile and not very base-rich waters, mostly in England. Deeper channels with consolidated pebbly beds are favoured and, *R. fluitans* being winter-green, quite large clumps of shoots can persist from year to year.

More strictly associated with calcareous waters in the limestone catchments of lowland England and Wales is the *Ranunculus penicillatus* var. *pseudofluitans* community (A17). Showing prolific growth where conditions are favourable, this vegetation can choke quite substantial rivers and is a major problem for flood control in some catchments.

It is sensible also to include here such starwort vegetation as we have sampled although, in phytosociological schemes, this is usually separated off from crowfoot communities into a fourth alliance of the Potametea, the Callitricho-Batrachion. We have recognised only a single assemblage from our limited data, the *Callitriche stagnalis* community (A16) which subsumes stands where other starworts can be dominant, *C. platycarpa* and, in southern Britain, *C. obtusangula*. This kind of vegetation can be found in shallow standing and sluggish waters and on their muddy margins but is most characteristic of fast to very swift, often spatey, waters in small sandy or gravelly streams.

Hairgrass and quillwort communities

Some of the vegetation types traditionally included among the hairgrass assemblages of the Littorelletea have already been dealt with among the mires of Volume 2 – the *Hypericum-Potamogeton* community of bog soakways (M29) and related vegetation of seasonally-inundated habitats (M30). These are best grouped together in the Hydrocotylo-Baldellion alliance.

Among the aquatics here, we have recognised two related communities. The *Littorella uniflora-Lobelia dortmanna* community (A22) includes open or closed swards of submerged or temporarily emergent vegetation dominated by mixtures of *Littorella* and *Lobelia* with tangles of *Juncus bulbosus* and occasional *Eleocharis palustris* and *E. multicaulis*. This community, which provides a locus for varieties such as *Elatine hexandra*, *Eleocharis acicularis*, *Subularia aquatica* and *Eriocaulon septangulare*, is characteristic of the barren, stony shallows of clear and infertile standing waters in upland lakes and pools.

Myriophyllum alterniflorum and *Isoetes lacustris* can both figure in the *Littorella-Lobelia* community where it extends into deeper and less turbulent waters, but we have characterised a separate *Isoetes lacustris/setacea* community (A23) where one or other of these quillworts, occasionally both together, dominate the sward in often dense stands or patches in our less fertile upland lakes. This kind of vegetation has sometimes been grouped with the *Littorella-Lobelia* community in the Lobelion dortmanna or separated off into another alliance, the Isoetion lacustris. Both assemblages are characteristic of the NCC Standing Water Site types 2 and 3.

Sphagnum-bladderwort vegetation

The bulk of the aquatic *Sphagnum* vegetation of highly acidic, nutrient-poor bog pools has been included in the NVC scheme among the Rhynchosporion communities described with the mires of Volume 2. Retained here, however, are free-floating mixtures of *Sphagnum auriculatum* and various bladderworts, what has usually been recorded as *Utricularia vulgaris* but what is actually more likely to be *U. neglecta*, together with *U. minor*, *U. intermedia* and perhaps also *U. ochroleuca*. Festoons of bulbous rush are a constant feature of the samples available from this vegetation which has accordingly been named as the *Juncus bulbosus* community (A24). It is characteristic of shallow base-poor, oligotrophic or dystrophic and often peaty waters in sheltered bays of lakes and small pools (NCC Standing Water Site types 1 and 2: Palmer *et al.* 1992) and would now generally be placed in a distinct alliance, the Sphagno-Utricularion, in the Utricularietea intermedio-minoris.

KEY TO AQUATIC COMMUNITIES

Much aquatic vegetation is relatively poor in species and many of the communities we have defined are overwhelmingly dominated by one or two taxa. In this case, in place of the usual dichotomous key or branching hierarchy of questions used elsewhere in the scheme, we have devised an alternative aid to identification of assemblages based on a synoptic table of species represented in the various vegetation types.

All sub-communities have been included in the table and those species diagnostic at frequency II or more rearranged in alphabetical order (Figure 5). As always, because the major distinctions between the communities and sub-communities are based on interstand frequency, this tabular key will work best where a number of samples of similar composition have been collected and made into a constancy table. It is the frequency values in this which should then be used to check for overall similarity with the vegetation types in the table.

Samples should always be taken from homogeneous stands and by 2 × 2 m or 4 × 4 m according to the scale of the vegetation or, where stands are narrow or of irregular form, of identical area but different shape. Small stands can be sampled in their entirety unless the vegetation is very fragmentary.

	1	2a	2b	2c	3	4	5a	5b	7a	7b	8a	8b	8c	8d	9a	9b	9c	10	11a	11b	11c	12	13a	13b	14	15	16a	16b	17	18	19	20	21	22a	22b	23	24a	24b
Apium nodiflorum	–	–																		–																		
Berula erecta		–	–		IV	=														–									=			–				–	–	
Callitriche hamulata																				–				=								–						
Callitriche hermaphroditica												=								–			=	=				=	=	III		–						=
Callitriche obtusangula					V							=	–										–	–			=	–	=	–	=							
Callitriche platycarpa		–			III	=	III	V		=	IV	=	=						–	–			–	–	=	–	=	–	–		=			=	>	=		
Callitriche stagnalis	–		–			V	V												–						=					=								
Ceratophyllum demersum					III	IV	IV												II	>			IV	III	–									=	>	=	III	=
Ceratophyllum submersum						=	=												=	III			–							=							=	
Chara spp.	–									=	=	=	=	=				=	–								=					–			=			
Elodea canadensis			IV	IV	IV	III	IV	IV	=		III	III	III	=				=	–	II	III		=	III	III	=	=	=	=	III	=	–		III	III	–	III	III
Elodea nuttallii						III	III				–	–							=	II	=		–	–	=												=	
Equisetum fluviatile		IV		IV										=																								
Glyceria fluitans	–							=			=								–							>	=		–									–
Glyceria × pedicellata							III				=								–								=	=										
Hippuris vulgaris	–											IV							–		II	II	=	=					=									=
Hottonia palustris														=																								
Hydrocharis morsus-ranae	–																		II	–			III	II	=		=	=		–		>		=	>		–	
Isoetes setacea																	=		=	=			–												=	>		
Isoetes lacustris																							–										–	=	III	=	–	–
Juncus bulbosus							II		>				=	=	=			=	II				=	=		>						>	II	=	III	=	>	>
Lemna gibba	>	>	>		>	V	III	=	–		–	=	–	–					–	–		=	=	–	–		=	–		–					–			–
Lemna minor	=	V	V	II	V	V	V	V	>	>	III	III	III				=	=	III	=	=	=	=	=	=	=	=	III	=	III	III	–	=	III	=	–	III	–
Lemna polyrhiza	=	V	=		>	V	V				=	–	–	=				=	–	–	–		–	–	–		=	=		=	=		=	–	–	–	–	–
Lemna trisulca	=		=		>	V						–	–						II	=	=		>	>	>							>	I	–	–	=	–	
Littorella uniflora																					=		>	>	=								>	>	>	>		=
Lobelia dortmanna																		>					–	–									>	≥	–	>		
Myriophyllum alterniflorum	–					=						–	–				=	>	>	II	III	>	>	>	=		=		=	II	=		=	–	I	III	=	
Myriophyllum spicatum						V						III	=	=	>	>	=		II	II	≥	>	>	>	>		>			=				III	>			
Myriophyllum verticillatum												=	–						II	II	=	>	=	=	=		–			=				III		=	–	
Nitella spp.	=			III	III												=		–	–			=	=	–						–			=	=		=	=
Nuphar lutea	=	–	–		=	III	=				–	III	–						III	III	III		=	III	=	=	=	–		III	III	–		III	III	=	III	–
Nymphaea alba	=														>	>	>		III	=			=	–	=					–					–	=		
Oenanthe aquatica															>	>	>		–	–	–							>										
Polygonum amphibium			–	–		IV							=	≡				>	III	III	–		>	–	=	=								=	>			=
Potamogeton berchtoldii						IV									>	>	>		II	=			≥	≥	–									III	>			
Potamogeton crispus	=		=	=		III	=	=			=	=	=	=					II	II	=		=	–	=			–					=				–	
Potamogeton filiformis	=																		II	=			=	–	=												=	
Potamogeton gramineus		–					=						=	≡				>	III	III			>	–	=									=	>		–	=
Potamogeton natans													–	–	>	>	>	–	–	–			≥	≥	–									=	≥	=		
Potamogeton obtusifolius					III	=						=	–				=	=	III	III	>		≥	=	=									=	–		III	=
Potamogeton pectinatus							=		≥			=	=					=	III	III	>		>	>	>		>							=		=	=	=
Potamogeton perfoliatus							=				=	–	–						II	II	=		–	–	=									=	–	=	–	
Potamogeton polygonifolius									≥												=		>	–	–			>						–	≡	=		
Potamogeton pusillus																			II	III	>		≥	–	=										III			
Potentilla palustris					=																		=	=	=			=							=			=

Ranunculus aquatilis	.	.	.	.	.	.	.	.	.	.	.	.	.	.	.	.	.	.	.	.	.	.	I	I	.	.	.	.	.	.	V	.	.	.	.	.	.	.
Ranunculus baudotii	.	.	.	.	.	.	.	.	.	.	.	.	.	.	.	.	.	.	.	I	III	.	.	I	.	.	.	.	.	.	.	.	V	.	.	.	.	.
Ranunculus circinatus	.	.	.	.	I	.	IV	.	.	.	.	I	I	.	.	.	.	.	I	II	I	.	.	.	.	I	.	.	.	.	.	I	.	.	.	.	.	.
Ranunculus fluitans	.	.	.	.	.	.	.	.	.	.	.	.	.	.	.	.	.	.	.	.	.	.	.	.	.	.	.	.	.	V	I	.	.	.	.	.	.	.
Ranunculus penicillatus	.	.	.	.	.	.	.	.	.	.	.	.	.	.	.	.	.	.	.	.	.	.	.	.	.	I	.	.	V	.	.	.	.	.	.	.	.	.
Ranunculus peltatus	.	.	I	.	.	.	.	.	.	.	.	.	.	.	.	.	.	.	I	I	I	I	I	I	I	.	.	.	.	.	.	V	.	.	.	.	.	.
Ruppia maritima	.	.	.	.	.	.	.	.	.	.	.	.	.	.	.	.	.	.	.	.	.	I	.	.	.	.	.	.	.	.	.	.	II	.	.	.	.	.
Sium latifolium	.	.	.	.	.	II	.	.	.	.	.	.	.	.	.	.	.	.	.	.	.	.	.	.	.	.	.	.	.	.	.	.	.	.	.	.	.	.
Sparganium angustifolium	.	.	.	.	.	.	.	.	.	II	.	.	.	.	.	.	I	.	.	.	.	.	I	I	.	.	.	.	.	.	.	.	.	.	.	I	III	I
Sparganium emersum	.	.	.	.	.	I	II	.	.	.	.	I	.	.	.	.	.	.	II	II	I	.	I	I	.	I	.	.	.	.	.	.	I	.	.	.	.	.
Sparganium erectum	.	.	.	.	.	II	.	.	.	.	.	.	.	.	.	I	.	.	.	.	.	.	.	.	.	.	.	.	.	.	.	I	.	.	.	.	.	.
Sphagnum auriculatum	.	.	.	.	.	.	.	.	.	.	.	.	.	.	.	.	.	.	.	.	.	.	.	.	.	.	.	.	.	.	.	.	.	.	.	.	I	IV
Stratiotes aloides	.	.	.	.	.	V	.	.	.	.	.	.	.	.	.	.	.	.	.	I	.	.	.	.	.	.	.	.	.	.	.	.	.	.	.	.	.	.
Utriculana vulgaris	.	.	.	.	.	IV	.	.	I	II	.	.	.	.	.	.	I	.	I	I	I	.	I	I	.	.	.	.	.	.	.	.	.	.	I	I	IV	I
Veronica beccabunga	.	.	.	.	.	.	.	.	.	.	I	.	.	.	.	.	.	.	.	.	.	.	.	.	.	.	.	.	II	.	.	.	.	.	.	.	.	.
Wolffia arrhiza	I	.	.	.	II	.	.	.	.	.	.	.	.	.	.	.	.	.	.	.	.	.	.	.	.	.	.	.	.	.	.	.	.	.	.	.	.	.
Zannichelia palustns	I	.	.	.	I	.	.	.	.	.	I	III	I	.	.	.	.	.	I	I	II	I	.	I	.	.	.	.	.	.	.	.	II	.	.	.	.	.

1 *Lemnetum gibbae*
2a *Lemnetum minoris*, Typical sub-community
2b *Lemnetum minoris*, *Lemna trisulca* sub-community
2c *Lemnetum minoris*, *Riccia fluitans-Ricciocarpus natans* sub-community
3 *Spirodela polyrhiza-Hydrocharis morsus-ranae* community
4 *Hydrocharis morsus-ranae-Stratiotes aloides* community
5a *Ceratophylletum demersi*, *Ranunculus circinatus* sub-community
5b *Ceratophylletum demersi*, *Lemna minor* sub-community
7a *Nymphaetum albae*, Species-poor sub-community
7b *Nymphaetum albae*, *Juncus bulbosus-Potamogeton polygonifolius* sub-community
8a *Nuphar lutea* community, Species-poor sub-community
8b *Nuphar lutea* community, *Callitriche stagnalis-Zannichellia palustris* sub-community
8c *Nuphar lutea* community, *Nymphaea alba* sub-community
8d *Nuphar lutea* community, *Potemogeton obtusifolius-Juncus bulbosus* sub-community
9a *Potamogeton natans* community, Species-poor sub-community
9b *Potamogeton natans* community, *Elodea canadensis* sub-community
9c *Potamogeton natans* community, *Juncus bulbosus-Myriophyllum alterniflorum* sub-community
10 *Polygonum amphibium* community
11a *Potamogeton pectinatus-Myriophyllum spicatum* community, *Potamogeton pusillus* sub-community
11b *Potamogeton pectinatus-Myriophyllum spicatum* community, *Elodea canadensis* sub-community
11c *Potamogeton pectinatus-Myriophyllum spicatum* community, *Potamogeton filiformis* sub-community
12 *Potamogeton pectinatus* community
13a *Potamogeton perfoliatus-Myriophyllum alterniflorum* community, *Potamogeton berchtoldii* sub-community
13b *Potamogeton perfoliatus-Myriophyllum alterniflorum* community, *Potamogeton filiformis* sub-community
14 *Myriophylletum alterniflori*
15 *Elodea canadensis* community
16a *Callitriche stagnalis* community, *Callitriche* spp. sub-community
16b *Callitriche stagnalis* community, *Potamogeton pectinatus* sub-community
17 *Ranunculus penicillatus* ssp. *pseudofluitans* community
18 *Ranunculetum fluitantis*
19 *Ranunculetum aquatilis*
20 *Ranunculetum peltati*
21 *Ranunculetum baudotii*
22a *Littorella uniflora-Lobelia dortmanna* community, *Littorella uniflora* sub-community
22b *Littorella uniflora-Lobelia dortmanna* community, *Myriophyllum alterniflorum* sub-community
23 *Isoetes lacustris/setacea* community
24a *Juncus bulbosus* community, *Ultricularia vulgaris* agg. sub-community
24b *Juncus bulbosus* community, *Sphagnum auriculatum* sub-community

Figure 5. Synoptic key to aquatic communities and sub-communities.

Floristic table A3

Spirodela polyrhiza	V (2–8)
Hydrocharis morsus-ranae	V (3–8)
Lemna minor	V (3–10)
Lemna gibba	V (1–8)
Lemna trisulca	V (2–8)
Berula erecta	IV (2–3)
Glyceria fluitans	IV (1–3)
Elodea canadensis	IV (3–9)
Ceratophyllum demersum	III (3–4)
Wolffia arrhiza	II (4–5)
Ranunculus circinatus	I (3)
Callitriche platycarpa	I (3)
Potamogeton crispus	I (2)
Azolla filiculoides	I (2)
Zannichellia palustris	I (3)
Nasturtium officinale	I (5)
Polygonum amphibium	I (2)
Number of samples	8
Number of species/sample	8 (7–11)

A4
Hydrocharis morsus-ranae-Stratiotes aloides community

Synonymy
Stratiotes aloides vegetation Pallis 1911, Ellis 1939, Lambert & Jennings 1951, Lambert 1965, all *p.p.*; *Hydrocharitetum morsus-ranae* van Langendonck 1935 *p.p.*; *Hydrocharito-Stratiotetum* (van Langendonck 1935) Westhoff 1946 *p.p.*; *Ceratophyllum-Stratiotes* nodum Wheeler & Giller 1982.

Constant species
Callitriche platycarpa, Ceratophyllum demersum, Hydrocharis morsus-ranae, Lemna minor, L. trisulca, Polygonum amphibium, Stratiotes aloides, Utricularia vulgaris.

Rare species
Myriophyllum verticillatum, Stratiotes aloides.

Physiognomy
The *Hydrocharis morsus-ranae-Stratiotes aloides* community has a usually abundant submerged element comprising luxuriant masses of *Ceratophyllum demersum* with *Utricularia vulgaris* and quite often some of the rare *Myriophyllum verticillatum*. *Callitriche platycarpa* occurs very frequently in small amounts and there is commonly a little *Elodea canadensis* and occasionally some *Potamogeton obtusifolius* or the rare *Ceratophyllum submersum*. Another rarity, *Stratiotes aloides*, is now particularly associated with this kind of vegetation within its much contracted range, and its striking large rosettes of rigid leaves can be quite abundant here. They remain submerged for much of the time, rising to the surface at flowering, but reproduction is by offsets in this country, male plants being very scarce and fruit never set (Clapham *et al.* 1962).

Floating above this layer is a surface mat of *Hydrocharis morsus-ranae, Lemna minor, L. trisulca, Polygonum amphibium* and *Nuphar lutea*, often mixed in with an abundant algal scum, mainly *Rhizoclonium*. Occasionally there are some emergent shoots of *Potentilla palustris, Hottonia palustris, Berula erecta,* *Oenanthe aquatica, Sparganium erectum, Nasturtium officinale, Sium latifolium, Alisma plantago-aquatica* or *Sagittaria sagittifolia.*

Habitat
The *Hydrocharis-Stratiotes* community is a very local vegetation type of mesotrophic, calcareous standing waters. It seems now to be restricted to a few localities, mostly in Broadland, though similar vegetation appears to have been more common there are one time, and perhaps occurred scattered through eastern England in the past.

In their study of the Catfield and Irstead fens in the Ant valley in Norfolk, Wheeler & Giller (1982*b*) showed that this community was most characteristic of those parts of the closed internal dyke system nearer to the fen margins, where there was direct or indirect influx of land-drainage waters and a shift from the deep peat of the fen basin to a mineral substrate. Chemical analysis of both waters and sediments there revealed higher levels of inorganic nitrogen (mainly nitrate) than in the more central areas of the fens, and slightly, but significantly, greater concentrations of calcium and magnesium throughout the year. They thought it possible that the mineral substrates around the fen edge, where the sediments showed higher redox characteristics, could have some influence on the gradient of trophic state and alkalinity, but suggested that it might be fertiliser run-off from the surrounding agricultural land, much chemically-manured and limed for the production of arable crops, that helped maintain the greater fertility in a comparatively nutrient-poor system.

The precise ecological needs of *Stratiotes*, the most distinctive plant in this community, are unclear. However, the *Hydrocharito-Stratioteum* described from the Continent has been designated characteristic of eutrophic waters (van Donselaar 1961) and, although Wiegleb (1978) considered *Stratiotes* typical of situations poor in both nitrogen and phosphorus, van Wirdum (1979) noted that the plant seemed to grow best in relatively nutrient-rich places, such as those irrigated by run-off from fertilised fields. Indeed, Westhoff & den Held (1969) attributed the extension of the range of *Stratiotes* in The Netherlands to increased use of fertilisers.

There is the interesting possibility, then, that the decline of the plant is related to nutrient depletion. Wheeler & Giller (1982*b*) gathered anecdotal evidence to support a previously abundant distribution of *Stratiotes* through much of the internal dyke system on the Catfield–Irstead fens, as recently as 15 years earlier: so much so, that the plants had to be dredged out of the waterways to allow passage of boats carrying away the sedge harvest. The decline there was much later than the more general loss of macrophyte vegetation from the open Broads (Phillips *et al.* 1978) and, although many stands of *Stratiotes* (as described by Pallis (1911), for example) disappeared from such sites, a more precise cause than simple 'nutrient enrichment' may have been responsible. Certainly, there has been increased input from agricultural run-off, but various workers have suggested that it is a rise in phosphorus content that is particularly important for the development of dense phytoplankton blooms and the loss of macrophytes (Phillips 1977, Moss 1978, Moss *et al.* 1979), with sewage effluent being an important contributory cause (Moss *et al.* 1979). Interestingly, Wheeler & Giller (1982*b*) noted that a small turf pond, isolated within the internal drainage system of the fens, close to the land margin but receiving no drainage from it, had virtually no macrophytes. Unlike the neighbouring dykes and pools with the *Hydrocharis-Stratiotes* community, this pond received input from a nearby septic tank.

The decline of *Stratiotes* has been much more widespread than just within Broadland and, although its distinctive appearance has commended it for introduction as an ornamental aquatic, its native localities are now few and far between across its former recorded range in East Anglia, the Fens and the north-east and north-west Midlands (Perring & Walters 1962). Through Europe it has a Continental distribution (Matthews 1955) extending from mid-Sweden down to northern Italy (Godwin 1975), and has probably never occurred in this country outside the warmer lowlands: all records, past and present, occur within the 28 °C mean annual maximum isotherm (Conolly & Dahl 1970) and, even in those rare situations where both sexes of the plant occur in the warmest part of the range, seed is never set. Some other members of this assemblage, although largely confined to the south and east of Britain, do retain a wider distribution than *Stratiotes* and, where they occur together in suitably unpolluted waters, could be considered as fragments of this kind of vegetation. From the rather vague descriptions provided of the dense growths of *Stratiotes* and *C.*

demersum that once characterised now largely barren open waters in places like the Broads, it is very difficult to tell whether they occurred in this exact community. Clearly, there is a general similarity, although some species, like *Hydrocharis* for example, were apparently not important in the open Broads.

Zonation and succession

The *Hydrocharis-Stratiotes* community persists as a local aquatic element in the fragmentary and abbreviated open-water transitions of dykes and pools, passing landwards to swamp and fen, and being replaced by other aquatic vegetation with shifts in the trophic state and base richness of the waters.

In the kind of extended open-water transitions described by Pallis (1911) and Lambert & Jennings (1951), *Stratiotes* is shown growing beyond or persisting among emergent *Scirpetum lacustris*, *Typhetum angustifoliae* and *Typhetum latifoliae* swamps, and in the pools and dykes where it survives today, very narrow strips or clumps of such communities can be interposed between it and the banks, together with patches of other swamps like the *Phragmitetum*. Or, the open waters may give way sharply to fen vegetation, usually in Broadland of the *Peucedano-Phragmitetum*, the more species-poor *Phragmites-Eupatorium* fen or the eutrophic *Phragmites-Urtica* fen.

Where the waters became more enriched in nitrate-nitrogen, calcium and magnesium, Wheeler & Giller (1982*b*) showed that the dykes of the Catfield–Irstead fens supported a fairly rich kind of *Elodea canadensis* vegetation, from which *Stratiotes*, *Utricularia vulgaris* and *Myriophyllum verticillatum* were absent, but where floristic continuity was maintained through the floating element of the flora. Towards the other extreme, where the waters were impoverished, this element also persisted, though in a rather attenuated form, together with *M. verticillatum*, *Potamogeton obtusifolius* and *Nuphar lutea*, with *Utricularia vulgaris* usually the most abundant plant. In this scheme, such assemblages are best retained here as fragmentary stands.

In spatial terms, this environmental and floristic gradient was disposed through the dyke system, running from the fen margin to its centre.

Distribution

The *Hydrocharis-Stratiotes* community is now very local and mostly confined to Broadland.

Affinities

This community has been characterised almost entirely from the small data set of Wheeler & Giller (1982*b*) and more comprehensive sampling of vegetation with *Stratiotes* is needed before a satisfactory definition of this kind of assemblage in Britain can be given. Its relationship with the *Spirodela-Hydrocharis* community also needs clarifying because some Continental schemes would group both communities together in a *Hydrocharitetum* (as in Oberdorfer 1977), while others would prefer a narrow view of a *Hydrocharito-Stratiotetum* (as in Westhoff & den Held 1969). There is also some disagreement as to whether this kind of vegetation should be placed in a Lemnion minoris (Oberdorder 1977) or a Hydrocharition (Westhoff & den Held 1969, de Lange 1972).

Floristic table A4

Hydrocharis morsus-ranae	V (1–4)
Stratiotes aloides	V (1–7)
Lemna minor	V (1–3)
Lemna trisulca	V (1–5)
Callitriche platycarpa	V (1–3)
Polygonum amphibium	IV (1–3)
Utricularia vulgaris	IV (1–7)
Ceratophyllum demersum	IV (1–7)
Myriophyllum verticillatum	III (1–3)
Nuphar lutea	III (1–2)
Potamogeton obtusifolius	III (1–6)
Elodea canadensis	III (1–4)
Potentilla palustris	II (1–2)
Hottonia palustris	II (1–2)
Berula erecta	II (1)
Ceratophyllum submersum	II (1–6)
Oenanthe aquatica	II (1–3)
Sparganium erectum	II (1–3)
Nasturtium officinale	II (1–3)
Sium latifolium	II (1–3)
Alisma plantago-aquatica	I (1–3)
Nymphaea alba	I (1–3)
Ranunculus lingua	I (1–3)
Sagittaria sagittifolia	I (1–3)
Potamogeton crispus	I (1)
Sparganium emersum	I (3–4)
Number of samples	12
Number of species/sample	14 (11–19)

A11

Potamogeton pectinatus-Myriophyllum spicatum community

Synonymy

Slow-moving water community Tansley 1911; Submerged-leaf association Pallis 1911; Moderate current vegetation Butcher 1933; Submerged community Tansley 1939; *Potamogeton filiformis-Chara* sociation Spence 1964; *Potametum pectinato-perfoliati* Den Hartog & Segal 1964; *Potamogeton perfoliatus* Gesellschaft Oberdorfer 1977.

Constant species

Myriophyllum spicatum, Potamogeton pectinatus.

Rare species

Potamogeton filiformis, P. trichoides, Ranunculus baudotii.

Physiognomy

The *Potamogeton pectinatus-Myriophyllum spicatum* community includes most of the richer and more diverse pondweed vegetation in which *Potamogeton pectinatus* and *Myriophyllum spicatum* play a prominent role. Usually, both these plants are present, often in some abundance by mid-summer, but stands in which just one or the other occurs, together with the characteristic community associates, can also be placed here. *P. pectinatus* is particularly striking where the vegetation extends into more nutrient-rich, standing or slow-moving waters, where it can make rapid growth in early summer from its overwintering tuberous buds or fruits, forming bushy clumps up to 2 m or so long. *M. spicatum* can also grow luxuriantly in this community, and tends to perform rather better than *P. pectinatus* where the flow is somewhat faster or has occasional spates, so it is often the more abundant plant where the vegetation extends into the moderately swift reaches of rivers. *M. alterniflorum*, however, is very scarce here and only occurs where there is some influence from more base-poor and oligotrophic waters, where the community begins to grade to the *Potamogeton-M. alterniflorum* vegetation characteristic of such situations. The rare *M.*

verticillatum occurs very locally in stiller waters, mostly in eastern England.

Other *Potamogeton* spp. occur with varying frequency and abundance in the community and further sampling might allow the definition of additional vegetation types of this general kind in which various of these pondweeds occur with more consistent prominence. Comparison with earlier accounts, however, suggests that local dominance among these associates was once much more widespread than is now the case (Pallis 1911, Tansley 1911, Pearsall 1921, Butcher 1933). *P. pusillus*, a linear-leaved pondweed like *P. pectinatus*, though not such a common plant, remains fairly frequent throughout, particularly in standing and more slow-moving waters, but *P. crispus* and *P. lucens*, both broad-leaved species, seem to be only occasional these days. More striking still is the relative scarcity of *P. perfoliatus*, another broad-leaved pondweed, which appears to be nothing like so common as previously, though, as with these other species, it can still be very abundant in some stands, occasionally to the exclusion of *P. pectinatus*.

P. natans sometimes provides a floating-leaved element to the vegetation, and there is occasionally some *P. berchtoldii* and, less often, *P. obtusifolius, P. praelongus, P. × nitens, P. alpinus* or the rare *P. trichoides* or *P. compressus*. Another rarity, *P. filiformis*, is particularly characteristic of one sub-community in shallower waters in Scotland, and in this kind of vegetation, too, *P. friesii* and *P. gramineus* have occasionally been recorded.

Locally luxuriant growth of these species can also be accommodated within the broad compass of the community, but just as commonly now it is the abundance of *Elodea* spp. among the *P. pectinatus* and *M. spicatum* that is the most striking feature of the vegetation. The commonest of these American introductions, *E. canadensis*, first recorded with certainty in Britain in the 1840s, is the most widespread here, varying from just occasional to very frequent in the different sub-communities and often growing very densely, but *E. nuttallii* can also occur in considerable quantity. Not distinguished in this country until 1966, this plant has now been found quite widely in England, occasionally in Wales and Scotland, and appears to be spreading, perhaps being able to oust *E. canadensis* under certain conditions (Simpson 1984, 1986). It has certainly been recorded preferentially here in one particular kind of *Potamogeton-M. spicatum* vegetation, where its more widespread relative is less frequent, but further sampling is needed to clarify where the ecological affinities of the plants are different, and much care needs to be taken to distinguish them because both show great phenotypic plasticity (Simpson 1988). Vegetation with the much rarer third species, *E. callitrichoides*, now thought extant in only two localities (Simpson 1986), has not been sampled.

Other plants frequent throughout this vegetation are few, but there is quite often some *Polygonum amphibium* and *Sparganium emersum*, and many stands have an abundance of *Chara* spp., though these have not been distinguished from one another in the sampling. *Nitella* spp. are much scarcer, though can show local prominence in some places. Occasionals include *Callitriche stagnalis*, *C. hermaphroditica*, *C. hamulata*, *C. obtusangula*, *Ranunculus peltatus*, *R. trichophyllus*, *Zannichellia palustris*, *Ceratophyllum demersum*, *Utricularia vulgaris*, *Juncus bulbosus*, *Apium inundatum* and the moss *Fontinalis antipyretica*. *Nuphar lutea*, *Nymphaea alba* and, more locally, *Nymphoides peltata* can occur as an open cover of floating-leaved vegetation, sometimes with *Lemna minor* and *Hydrocharis morsus-ranae*, and shoots of *Glyceria fluitans* or *G. plicata* can trail into shallows from water margins.

Sub-communities

Potamogeton pusillus sub-community. Mixtures of *P. pectinatus* and *M. spicatum*, together with the preferential *P. pusillus* and *E. nuttallii*, form the bulk of the cover here, with *Chara* spp. also rather more frequent than elsewhere in the community, and *P. lucens* occurring occasionally. *E. canadensis*, *Sparganium emersum* and *Polygonum amphibium* are all fairly common and there is sometimes a little *P. crispus*, *P. trichoides* and *Ranunculus circinatus*.

Elodea canadensis sub-community. *P. pusillus* remains fairly frequent in this sub-community, but along with *P. pectinatus* the commonest pondweeds are usually *P. crispus* and *P. natans* with more occasional *P. berchtoldii*, *P. perfoliatus* and *P. obtusifolius*. *M. spicatum* is still constant; like each of the *Potamogeton* spp., it can grow luxuriantly here. Very often, though, *E. canadensis* is the most abundant plant, only rarely with, or replaced by, *E. nuttallii*. There is sometimes a lot of *Lemna trisulca*

entangled among the pondweeds and *Ranunculus circinatus* occurs rather more often here than elsewhere in the community. Very occasional, but confined to this kind of *Potamogeton-M. spicatum* vegetation, are the rarities *P. compressus*, occurring scattered through the Midlands and East Anglia, *Stratiotes aloides* in eastern England, *Luronium natans* towards the north-west of England and in Wales, and the annual *Elatine hydropiper*, found in just a very few localities across the country.

Among the community associates, *Polygonum amphibium* and *Sparganium emersum* are fairly frequent, with occasional *Zannichellia*, *Juncus bulbosus*, *Utricularia vulgaris*, *Callitriche stagnalis*, *C. hamulata*, *Ceratophyllum demersum* and *Glyceria fluitans*.

Potamogeton filiformis sub-community: *Potamogeton filiformis-Chara* sociation   Spence 1964. Both *P. pectinatus* and *M. spicatum* are constant here and many stands have an abundance of *Chara* spp. Particularly distinctive, however, is the frequency and local prominence of the rare *P. filiformis*, especially abundant in shallow waters, with occasional *P. friesii*, *P. gramineus* and *P. polygonifolius* extending deeper. *P. pusillus* remains quite common in waters of moderate depth and, among other pondweeds of the community, *P. berchtoldii*, *P. perfoliatus*, *P. obtusifolius*, *P. crispus* and *P. natans* all occur quite often and each can show local dominance, particularly in deeper waters. *Elodea* spp. are poorly represented but there is very often some floating-leaved *Polygonum amphibium*.

Other preferential features of this vegetation are the high frequency of *Hippuris vulgaris* and *Littorella uniflora* with common *Ranunculus baudotii* and occasional *R. trichophyllus*, *Callitriche stagnalis*, *C. hermaphroditica* and *Zannichellia palustris*. *Myriophyllum alterniflorum* is confined to this kind of *Potamogeton-M. spicatum* vegetation, and there can be sparse records for *Lobelia dortmanna* or *Ruppia spiralis*. Among the general community associates, *Apium inundatum*, *Fontinalis antipyretica* and *Juncus bulbosus* are fairly common.

Habitat

The *Potamogeton-M. spicatum* community is best developed on finer mineral substrates in clear, standing to moderately fast-moving waters, which are generally mesotrophic to eutrophic and base-rich, sometimes marly, occasionally slightly saline. It occurs widely throughout Britain, mostly in the lowlands, being found in dykes, canals, ponds and lakes, where it can penetrate to considerable depths, and extending into streams and rivers where these are not too swift or spatey. This kind of aquatic vegetation can stand a modest amount of eutrophication, but it has become increasingly impoverished and local with the greater pollution and turbidity of many sites.

The community subsumes the bulk of our more diverse *Potamogeton* vegetation in fertile and base-rich waters. The pH and alkalinity vary somewhat among the different sub-communities, but are generally between 7 and 8.5 with 30–125 mg l⁻¹ calcium carbonate and conductivities of 200–1000 μmho (Palmer 1992, Palmer *et al.* 1992). Such conditions are quite widely met through the lowlands of Britain in waters occurring on, or fed by drainage from, less acidic rocks and superficials and rich examples of this kind of vegetation were described in early studies from the Norfolk Broads and streams and rivers flowing from the East Anglian Chalk (Pallis 1911, Tansley 1911, Butcher 1933). Stands have been noted subsequently in many open waters in England, including artificial examples like the Tring Reservoirs in Hertfordshire, the Chew Valley and Blagdon reservoirs in Somerset and the flooded gravel pits of the Cotswold Water Park, and the sluggish to moderately fast-flowing reaches of rivers such as the Hampshire Avon and the Wye (Ratcliffe 1977). The community also occurs in some stretches of canal, like the Prees Branch of the Shropshire Union, and in dykes, though here stands are often fragmentary (e.g. Pevensey Levels Survey 1986). Towards the west and north, suitable sites become scarcer, though the local occurrence of lakes on more calcareous drift, as at Llangorse in the Brecon Beacons and Hornsea Mere in Yorkshire, or on limestone, as at Malham Tarn and over the Durness rocks of Sutherland, can provide congenial habitats, and the slower reaches of more eutrophic rivers such as the Tweed also have this vegetation (Ratcliffe 1977). Deposition of fine mineral material in waters that are not generally very fertile or base-rich can also create locally suitable conditions, as in the Cumbrian lakes described by Pearsall (1918, 1921). Particularly striking, too, along the north-western seaboard of Scotland are the sites provided for the community in open waters influenced by calcareous wind-blown sands. These do occur locally elsewhere in Britain, as at Kenfig Pool in Glamorgan, Llyn Coron in Anglesey and the Loch of Strathbeg in Aberdeenshire, but it is in the pools and lakes on the machair landscape of the Isles, notably South Uist, and at scattered mainland localities in Ross and Sutherland, that the stands are the most extensive and richly developed.

In these various kinds of sites, the richer and more luxuriant stands of the *Potamogeton-M. spicatum* community are confined to the quieter stretches of water with beds of finer mineral materials, away from the wave-disturbed zone on exposed shores in lakes and pools, and the swift or more spatey reaches of moving waters. In fact, the two constants of the community are somewhat more tolerant of turbulent conditions than

most of the other associates of this vegetation, and it is small individuals of these species which often characterise the open cover in more disturbed situations. *P. pectinatus* is quite susceptible to uprooting or damage to its shoots when the plants are small in the spring or where bushy growth experiences summer storm flows, but it readily regrows, and it will tolerate the moderately turbid conditions that can develop with bed disturbance. *M. spicatum* is shallow-rooted and likely to be dislodged with little force, but it can remain vigorous in faster flows than *P. pectinatus* and will survive long from battered fragments (Haslam 1978).

In less turbulent situations, various of the other *Potamogeton* spp. can make abundant growth in this community, sometimes forming the locally dense patches of one or two pondweeds that were described as consocies or sociations in earlier accounts (Pearsall 1921, Spence 1964). And, where there is some gradation in depth of the waters, these may be disposed, within the overall cover of the vegetation, in a zoned fashion. Species such as *P. filiformis* and, to a lesser extent, *P. pusillus* are confined to shallower situations, *P. pectinatus* extending further, *P. perfoliatus*, *P. gramineus*, *P. praelongus* and *P. obtusifolius* penetrating to the greatest depths reached by the community (Pearsall 1918, 1921, Tansley 1939, Spence 1964).

Very often now, though, such local patterning is overwhelmed by the general abundance throughout of *P. pectinatus* or quite commonly by dense growth of one or other of the *Elodea* spp., and both the *P. pusillus* and *Elodea* sub-communities contain many stands of this kind in which the associate species are reduced to sparse individuals. There is little doubt that such a change reflects in part the increasing eutrophication of open waters from such sources as agricultural run-off and sewage discharge. *P. pectinatus*, for example, is very tolerant of such enrichment and will actually grow more luxuriantly in moderately polluted and turbid waters, provided these are not too shallow or fast-moving (Haslam 1978). Richer waters which are relatively unpolluted, clear and sluggish or standing also provide a very congenial medium for the opportunistic expansion among this vegetation of *E. canadensis* (Haslam 1978, Simpson 1983, 1984). This appears to have spread through much of its existing range in Britain by the middle of the present century, though there is some evidence that it is still continuing its advance in less oligotrophic waters in northern Scotland (Simpson 1983). *E. nuttallii* remains much less widespread, though it seems to have very similar ecological requirements to *E. canadensis*. It is possible, however, that it fares the better in less nutrient-rich situations, and it may be able to oust *E. canadensis* where it invaded subsequently (Simpson 1983, 1984). Trophic differences may play some part in the floristic contrasts between the *P.*

pusillus and *Elodea* sub-communities, although the former also appears to be more characteristic of shallower and perhaps more sluggish waters than the latter.

These two kinds of *Potamogeton-M. spicatum* vegetation, particularly the *Elodea* sub-community, occur widely through lowland England, with less common occurrences in Wales and southern and eastern Scotland. The *P. filiformis* sub-community, on the other hand, is very much confined to the seaboard of Scotland, especially towards the west, largely beyond the present range of the *Elodea* spp., and best developed of all in the pools and lochs of the machair. Here, there is not only the phytogeographic peculiarity of the occurrence of the rare Continental Northern pondweed *P. filiformis*, but also a more frequent diversity among the abundant associates, particularly those tolerant of more calcareous conditions. In sites like lochs Bornish, Upper Kildonan, Torornish, Stilligarry and Grogarry on South Uist, for example, the pH is often at the highest levels associated with the community, frequently over 8, with conductivities often over 500 μmho (Palmer 1992, Palmer *et al.* 1992). The waters are generally clear and often marly, and in exposed situations there is the added complication of some saline influence from salt-laden winds or percolating ground waters. There is often considerable variation within and between stands in the abundance of the plants, and zonations in the deepening waters over shelving shores are more frequently evident here than in the other sub-communities. *P. filiformis* commonly dominates in the shallows, *P. pectinatus* and *P. gramineus* becoming prominent beyond this, *P. obtusifolius*, *P. perfoliatus* and *P. praelongus* usually penetrating furthest, to 3 m or so where the waters are particularly clear and still. This kind of vegetation also provides a frequent locus for *Hippurus vulgaris* and *Ranunculus baudotii* in some of their few Scottish localities, the former reaching depths of 1 m or more in extreme cases.

Zonation and succession
The *Potamogeton-M. spicatum* community is found alone, or with other submerged aquatic vegetation, extending to the limits of colonisation in less turbulent, base-rich waters, thinning out in more exposed situations and being replaced by other communities where conditions become less calcareous and eutrophic. Open covers of floating or floating-leaved aquatics can be found on the surface above the community and in shallower waters there can be invasion by swamp vegetation.

Through the lowlands of southern and eastern Britain, the *P. pusillus* and *Elodea* sub-communities can be found in less polluted, sluggish waters grading to locally dense stands of *Elodea* vegetation in which associate species are very few and sparse, or in shaded situations

to luxuriant patches of the *Ceratophylletum demersi*. Very commonly, too, especially with the *Elodea* sub-community, there is an associated floating mat of the *Lemnetum minoris* or sometimes an open cover of the *Polygonum amphibium*, *Potamogeton natans* or *Nuphar lutea* floating-leaved vegetation. The most frequently associated swamps of open-water transitions from such aquatic assemblages are the *Sparganietum erecti*, the *Typhetum latifoliae*, the *Phragmitetum* and the *Glycerietum maximae*, with the *Caricetum ripariae* and *Scirpetum lacustris* more occasionally represented. Around the shelving water margins of small pools, dykes and canals and along the banks of streams, various kinds of Glycerio-Sparganion vegetation can also be found trailing out into the *Potamogeton-M. spicatum* community.

Where there is an increase in turbulence in such waters, shifting into wave-disturbed zones or into more fast-moving or spatey streams, the *P. pusillus* and *Elodea* sub-communities become increasingly sparse in cover, and impoverished in their species content, such that just scattered, puny plants of *P. pectinatus* and *M. spicatum* finally remain. Further upstream in base-rich moving waters, these kinds of vegetation can be replaced by the *Ranunculetum fluitantis*, the *Ranunculus penicillatus* community or *Callitriche* stands.

Patterns involving the *P. filiformis* sub-community in its distinctive north-west Scottish sites are of a rather different character, particularly where there is some influence within the same loch or pool of peaty, base-poor waters, quite a common and very striking feature over parts of the machair. Here, such submerged vegetation as the *Elodea* communities and the *Ceratophylletum demersi* are rarely present, although where there is some saline influence, the *Potamogeton-M. spicatum* vegetation may give way to the *Ruppietum maritimae* in which such species as *P. pectinatus*, *Ranunculus baudotii*, *Hippuris vulgaris* and *Zannichellia palustris* maintain some representation. More often, with the decreasing influence of shell-sand, there is a fall in pH and conductivity of the waters and the community is replaced by the *Potamogeton-M. alterniflorum* vegetation. The *P. filiformis* sub-community can be seen as transitional to this assemblage and, indeed, *P. pectinatus*, *P. filiformis*, *P. pusillus* and *P. natans* can run on into it at moderately high frequency and with some local abundance. Typically, though, it is *P. perfoliatus* and *P. gramineus* which provide the most consistent element of the less calcicolous vegetation, together with *M. alterniflorum*, which almost completely replaces *M. spicatum*. In some places, where peaty shallows pass quite suddenly to shelly beds further out, the switch from one vegetation type to the other is quite sharp, but often the influence of the different substrates is less well defined, when transitions can be extensive. There is often, too, some *Littorella uniflora* amongst both assemblages, which adds to the continuity of the pattern, and this can thicken up locally in zonations to the *Littorella-Lobelia* community, especially where the substrate becomes gravelly. A floating-leaved element may be represented by the *Nymphaeetum albae* or the *Potamogeton natans* community and the usual swamps associated with the margins of these waters are the *Eleocharitetum palustris*, the *Equisetetum fluviatilis*, the *Caricetum rostratae* or, occasionally, the *Phragmitetum*.

Such sites have, for the most part, been spared any marked influence of eutrophication or pollution but, elsewhere, the *P. pusillus* and *Elodea* sub-communities have often been replaced by impoverished *P. pectinatus* vegetation with increased richness and contamination of the waters.

Distribution
This kind of vegetation is widespread but local throughout lowland England with more occasional stands in Wales and Scotland. The *P. filiformis* sub-community is largely confined to the western seaboard of Scotland, with the other types occurring through the rest of the range.

Affinities
The *Potamogeton-M. spicatum* community subsumes most of the richer *Potamogeton* vegetation of our most base-rich waters. It is clearly less diverse than some of the pondweed assemblages described in early accounts (Pallis 1911, Tansley 1911, 1939, Butcher 1933), but this is in part a real reflection of the changes that have occurred among this kind of aquatic vegetation in recent years. Further sampling is needed to see whether a range of discrete communities could be characterised from among the variation that is included here, and to relate this kind of vegetation to associations described from the Continent by den Hartog & Segal (1964) and Oberdorfer (1977) and now generally placed in the Parvopotamion alliance.

Floristic table A11

	a	b	c	11
Potamogeton pectinatus	IV (1–10)	IV (1–10)	IV (1–10)	IV (1–10)
Myriophyllum spicatum	IV (1–8)	IV (1–8)	IV (1–10)	IV (1–10)
Chara spp.	IV (1–8)	II (1–10)	III (1–10)	III (1–10)
Potamogeton pusillus	IV (1–10)	II (1–10)	II (1–10)	II (1–10)
Elodea nuttallii	III (1–10)	I (1–10)		I (1–10)
Potamogeton lucens	II (1–10)	I (1–6)		I (1–10)
Elodea canadensis	II (4–8)	IV (1–10)	I (4–10)	III (1–10)
Potamogeton crispus	I (4–6)	III (1–10)	II (1–10)	II (1–10)
Potamogeton natans	I (1–5)	III (1–10)	II (1–5)	II (1–10)
Lemna trisulca		II (1–10)	I (1–8)	I (1–10)
Ranunculus circinatus	I (1–8)	II (1–10)	I (4)	I (1–10)
Elatine hydropiper		I (1–8)		I (1–8)
Luronium natans		I (8–10)		I (8–10)
Stratiotes aloides		I (4–8)		I (4–8)
Potamogeton compressus		I (1–10)		I (1–10)
Potamogeton filiformis		I (10)	IV (1–8)	II (1–10)
Hippuris vulgaris		I (4–10)	IV (1–5)	II (1–10)
Ranunculus baudotii		I (4–6)	III (1–10)	I (1–10)
Potamogeton friesii	I (4)	I (4–10)	II (1–10)	I (1–10)
Zannichellia palustris	I (4)	I (1–10)	II (1–10)	I (1–10)
Littorella uniflora	I (6)	I (4–6)	III (4–10)	II (4–10)
Callitriche stagnalis	I (4)	I (1–8)	II (1–8)	I (1–8)
Callitriche hermaphroditica		I (10)	II (4–8)	I (4–10)
Ranunculus trichophyllus		I (4–10)	II (1–10)	I (1–10)
Potamogeton gramineus		I (1)	II (1–10)	I (1–10)

Floristic table A11 (*cont.*)

	a	b	c	11
Myriophyllum alterniflorum			I (1–8)	I (1–8)
Lobelia dortmanna			I (4–6)	I (4–6)
Ruppia spiralis			I (1–5)	I (1-5)
Polygonum amphibium	II (1–4)	III (1–10)	III (1–10)	III (1–10)
Sparganium emersum	II (1–4)	II (1–5)	I (1–4)	II (1–5)
Potamogeton berchtoldii	I (1–8)	II (1–10)	II (1–10)	II (1–10)
Potamogeton perfoliatus	I (1–10)	II (4–10)	II (1–10)	II (1–10)
Nitella spp.	I (8–10)	I (1–10)	I (1–10)	I (1–10)
Glyceria fluitans	I (1–4)	I (1–6)	I (1–8)	I (1–8)
Fontinalis antipyretica	I (1–6)	I (1–10)	I (1–10)	I (1–10)
Potamogeton obtusifolius	I (6)	I (1–10)	I (1–10)	I (1–10)
Nuphar lutea	I (1–4)	I (1–5)	I (1–4)	I (1–5)
Lemna minor	I (1–3)	I (1–4)	I (1–4)	I (1–4)
Potamogeton trichoides	I (1–4)	I (6–10)	I (1)	I (1–10)
Nymphaea alba	I (1–4)	I (2–4)	I (2–4)	I (1–4)
Callitriche hamulata	I (4)	I (1–10)	I (1)	I (1–10)
Utricularia vulgaris	I (8)	I (4–6)	I (1–4)	I (1–6)
Ranunculus peltatus	I (4)	I (6)	I (1–2)	I (1–6)
Ceratophyllum demersum	I (1–6)	I (4–10)	I (4)	I (1–10)
Callitriche platycarpa	I (4)	I (1–4)	I (4)	I (1–4)
Glyceria plicata	I (1–4)	I (1–4)		I (1–4)
Hydrocharis morsus-ranae	I (4)	I (4–8)		I (4–8)
Juncus bulbosus	I (4)	I (1–10)	I (4–10)	I (1–10)
Nymphoides peltata	I (8)	I (4)		I (4–8)
Potamogeton alpinus		I (1–6)	I (1–6)	I (1–6)
Apium inundatum		I (4–8)	I (1–4)	I (1–8)
Potamogeton praelongus		I (8)	I (1–8)	I (1–8)
Potamogeton × nitens	I (4)		I (4–6)	I (4–6)
Callitriche obtusangula		I (4)	I (4–6)	I (4–6)
Potamogeton polygonifolius		I (6)	I (1–4)	I (1–6)
Ranunculus omiophyllus		I (4)	I (1)	I (1–4)
Oenanthe aquatica		I (1)	I (1–4)	I (1–4)
Number of samples	74	73	100	247
Number of species/sample	7 (2–10)	10 (4–16)	8 (4–21)	8 (2–21)

a   *Potamogeton pusillus* sub-community
b   *Elodea canadensis* sub-community
c   *Potamogeton filiformis* sub-community
11   *Potamogeton pectinatus-Myriophyllum spicatum* community (total)

A12
Potamogeton pectinatus community

Synonymy
Potamogeton pectinatus-Gesellschaft Oberdorfer 1977.

Constant species
Potamogeton pectinatus.

Physiognomy
The *Potamogeton pectinatus* community comprises spe-cies-poor vegetation dominated by this fine-leaved pondweed. It overwinters as fruits or small tuberous buds, but can make rapid growth in early summer to form bushy clumps, up to 2 m or so long and often very luxuriant, among which there are but few and typically only sparse associates. Small patches of duckweed thalli, with *Lemna gibba* and/or *L. minor*, are sometimes found floating above the stands in quieter waters, but sub-merged there are generally just scattered individuals of such plants as *Callitriche stagnalis, Myriophyllum spica-tum, Elodea canadensis* and *Ceratophyllum demersum* with infrequent *E. nuttallii, L. trisulca* and *Zannichellia palustris.* In brackish water, the last can be joined by *Ruppia spiralis* and, in Scottish sea lochs, by *Fucus ceranoides.* Other *Potamogeton* spp. are rare, but *P. perfoliatus* has been recorded very occasionally.

Habitat
This kind of vegetation is characteristic of still to quite fast-moving, eutrophic waters, often with some measure of artificial enrichment, and frequently polluted and turbid. It is widespread through the warmer lowlands of Britain and has become increasingly common in pools and canals, dykes and streams as these have been contaminated by agricultural and industrial effluents and sewage.

P. pectinatus is a frequent plant through much of lowland England and the milder parts of the north and west, where the mean annual maximum temperature is for the most part above 25 °C (Conolly & Dahl 1970). Through these regions, it is characteristic of various kinds of aquatic vegetation in sites that are naturally rich in cations and nutrients, and sometimes attains the sort of overwhelming dominance typical of this community in such unpolluted situations. Vigorous stands can be found, for example, in clean, standing or sluggish waters with clay or silt beds, habitats once common enough in the subdued scenery over softer, sedimentary rocks in the south and east of the country. But *P. pectinatus* seems especially well able to capitalise on the eutrophication that is now so widespread in our intensive agricultural landscapes and around settle-ments and industrial developments. It is also tolerant of various kinds of chemical pollution, indeed will actually grow more luxuriantly at certain levels of contamination with some effluents, provided the waters are not too fast-moving, shallow or very turbid (Haslam 1978). Under such conditions, most other aquatics suffer a marked decline, and those which are able to persist somewhat better may be overwhelmed by the bushy growth of pondweed, leaving the sort of impoverished stands typical here. With very heavy pollution, however, even *P. pectinatus* fails and, in our most strongly contami-nated rivers and pools it is usually very sparse stands of this community that mark the limit of the vegetated reaches.

In cleaner waters, this kind of vegetation remains most common in the lower stretches of hill rivers to the west of the country and in the middle reaches of some chalk-clay streams but, in such moving waters, fast flow and spate exert another check on the vigour of *P. pectinatus.* Young plants are readily uprooted in their early weeks of spring growth and, though the deep and extensive rhizome system that develops by the summer gives some protection against erosion from the bed, the shoots remain quite susceptible to turbulence and are easily damaged. Individuals can grow large in slow flows but, in more quick-moving waters, growth may be balanced by loss, and the species is intolerant of conti-nual fast flow or frequent spates (Haslam 1978).

In especially congenial conditions, this vegetation may become so luxuriant as to choke the flow in smaller water-courses, such that cutting becomes necessary. Two or even three cuts during the summer are sometimes needed, but one usually suffices and *P. pectinatus* will not recover if chopped back late in the season. The plant can become abundant after dredging early in the year and is moderately tolerant of the turbidity that disturbance produces (Haslam 1978).

Zonation and succession

In highly eutrophicated and polluted sites, this community may be the only kind of aquatic vegetation to survive but, in cleaner conditions, it can be found with various other submerged, floating and floating-leaved assemblages typical of richer waters.

In more base-rich pools, dykes and streams, the *P. pectinatus* community can sometimes be clearly seen as an impoverished form of the *Potamogeton-M. spicatum* vegetation, developed where this particular species of pondweed has attained local luxuriance, and grading to the richer assemblage with an increase in the frequency and cover of *M. spicatum*, *Elodea canadensis* or *E. nuttallii* and other *Potamogeton* spp., such as *P. perfoliatus*, *P. crispus*, *P. pusillus* and *P. lucens*. The *Elodea* spp. may themselves thicken up patchily into species-poor stands, and there can also be stretches among these mixtures dominated by the *Ceratophylletum demersi*. The *Polygonum amphibium* community may also form a canopy of floating leaves in the shallower waters, with *Nuphar lutea* stands occurring further out in deeper pools, dykes and slow-moving waters. Mats of the *Lemnetum gibbae* and the *Lemnetum minoris* may occur in sheltered situations. A variety of emergents can invade the margins of such open waters, with communities like the *Sparganietum erecti*, the *Typhetum latifoliae*, the *Phragmitetum* and the *Glycerietum maximae* developing as a zone in the shallows or as clumps distributed along the edges of dykes and streams. These may persist around the fringes of eutrophicated and polluted waters, where the aquatic element is reduced to clumps of the *P. pectinatus* community and sparse *Elodea* vegetation and *Ceratophylletum demersi*.

In brackish situations, the *P. pectinatus* community may again be the sole surviving vegetation where there has been some contamination of the waters but, in cleaner dykes and pools in, for example, coastal marshes, it can occur with the richer *Potamogeton-M. spicatum* assemblage and dense stands of the *Ceratophylletum submersi*. In this latter, *P. pectinatus* can maintain its constancy but it yields in abundance to plants such as *Ceratophyllum submersi*, *Zannichellia palustris* and *Ranunculus baudotii*. Floating mats of the *Lemnetum minoris* can also occur over such vegetation and there can be colonisation by the *Scirpetum maritimi* or *Scirpetum tabernaemontani*.

Fragmentary patterns of the kind described for fresh waters will persist in quite swift flows, as in mill races, watercress bed feed-streams and the upper reaches of rivers and, in such situations, there is often some zonation to patches of the *Callitriche stagnalis* community or various kinds of crowfoot vegetation. There can also be some seasonal variation in the prominence of the different dominants too, with *C. stagnalis* and the *Ranunculus* spp. showing luxuriance in late spring and early summer, *P. pectinatus* reaching its peak of abundance later in the year, along with the Glycerio-Sparganion vegetation that frequently trails into the margins of such waters. In chalk streams, some of these marginal plants, like *Apium nodiflorum* and *Nasturtium officinale*, together with various crowfoot taxa, can penetrate to the higher reaches, while the *Callitriche* and *P. pectinatus* communities peter out at lower levels. Along waters draining sandstone rocks and shales, *Callitriche* vegetation can replace the *P. pectinatus* community with the shift to swifter flows upstream.

Distribution

The community has a wide distribution through the lowlands of southern Britain, with more sporadic records in the west and north, and has become much commoner with the pollution of open waters.

Affinities

Vegetation of this kind, little described except in recent surveys of ditch aquatics, is best seen as an impoverished form of the sort of Parvopotamion assemblages included in the *Potamogeton-M. spicatum* community. Oberdorfer (1977) has characterised a comparable association from similar situations in Germany.

material, the *Myriophylletum* gives way to the *Potamogeton-M. alterniflorum* community. The two vegetation types intergrade, but there *M. alterniflorum* is typically accompanied by, not only *L. uniflora*, but also a variety of pondweeds, of which the commonest are *Potamogeton perfoliatus*, *P. gramineus*, *P. pusillus*, *P. berchtoldii* and *P. obtusifolius*, all of these more prone to battering than the milfoil. Dense stands of *Chara* or *Nitella* spp. also often occur. A clear zonation involving all these communities is shown in Pearsall's (1918) sketch of aquatics in Fold Yeat Bay in Esthwaite.

The *Potamogeton-M. alterniflorum* vegetation can also be seen replacing the *Myriophylletum* in moving waters where the flow becomes a little slacker or less spate-prone in moving downstream, and the deposition of mineral material more extensive. Towards the upper reaches of such streams, the *Myriophylletum* can show extensive overlap with the *Callitriche stagnalis* community, and indeed this vegetation may replace it where vascular plants retain a patchy presence far up into the torrent-like heads of such river systems. In other places, the *Myriophylletum* represents the limit of such vegetation before bryophyte communities of submerged rocks dominate the aquatic environment.

Another fairly frequent transition can be seen where quieter stretches of peaty waters interrupt or replace more fast-flowing rocky reaches with the shift to gentler, boggy terrain. Then, the *Myriophylletum* often gives way to *Potamogeton-Ranunculus* soakways, with *P. polygonifolius*, *Ranunculus flammula*, *R. omiophyllus* and *Agrostis stolonifera*. Pools along such trickling waters may also have *Juncus bulbosus* vegetation, but *M. alterniflorum* only rarely extends into such dystrophic assemblages.

Distribution

The *Myriophylletum* is widespread and fairly common through the upland fringes of the north and west of Britain, with very local occurrences in suitable habitats in the lowland south and east.

Affinities

The community is equivalent to the *M. alterniflorum*-dominated vegetation previously described from British waters (Pearsall 1918, Spence 1964), and from the Continent as the *Myriophylletum alterniflori* Lemée 1937 *emend.* Sissingh 1943. Westhoff & den Held (1969) located this assemblage in the Potamion graminei with other communities of nutrient-poor waters of the order Luronio-Potametalia. A simpler solution would be to place it with most of the other pondweed vegetation in the Parvopotamion.

Floristic table A14

Myriophyllum alterniflorum	V (1–10)
Juncus bulbosus	II (3–5)
Littorella uniflora	II (1–4)
Callitriche hamulata	II (2–5)
Fontinalis antipyretica	II (1–4)
Chara spp.	II (1–7)
Potamogeton natans	II (1–5)
Scirpus fluitans	I (6–8)
Nitella spp.	I (2–4)
Ranunculus peltatus	I (4–8)
Equisetum fluviatile	I (5)
Lobelia dortmanna	I (3)
Lemna minor	I (1)
Callitriche stagnalis	I (4)
Potamogeton filiformis	I (1–3)
Potamogeton gramineus	I (1–3)
Number of samples	16
Number of species/sample	3 (1–6)

A15
Elodea canadensis community

Synonymy
Elodetum Matthews 1914, Tansley 1939; Medium-slow current vegetation Butcher 1933 *p.p.*; Pflanzengesellschaft mit *Elodea canadensis* Solinska 1963; *Elodeetum canadensis* (Pignatti 1953) Passarge 1964; *Elodea* society Spence 1964.

Constant species
Elodea canadensis.

Physiognomy
The *Elodea canadensis* community comprises species-poor vegetation in which this particular kind of North American pondweed has become dominant. The plant overwinters as short unbranched stems or as turions but, by early summer, it can become very abundant, either as free-floating masses or with the shoots rather loosely anchored in the substrate. In such stands, other species are at most only occasional, though quite a variety are represented at low frequency. Commonest among other submerged aquatics are *Ceratophyllum demersum*, *Myriophyllum spicatum*, *Potamogeton pectinatus*, *P. crispus*, *P. perfoliatus*, *Callitriche stagnalis*, *C. hamulata*, *Ranunculus circinatus* and *R. penicillatus* ssp. *pseudofluitans*. Then, there are quite often some associated mats of duckweeds *Lemna minor* and *L. gibba*, with *Potamogeton natans* and *Polygonum amphibium* forming an open canopy of floating leaves. *Glyceria fluitans* shoots sometimes trail in from water margins.

Habitat
The *Elodea canadensis* community is most characteristic of still to sluggish, nutrient-rich waters, shallow to quite deep, and generally with fine mineral beds. Since the introduction of the plant, some century and a half a go, it has become very widespread through the warmer lowlands of Britain, occurring commonly in ponds and lakes, canals, dykes and slow-moving rivers and streams. Although its overall range seems to have now stabilised, *E. canadensis* will still actively colonise new water bodies, and the community often makes an early appearance in sites like flooded gravel pits, as well as sometimes quickly returning to waters recovering from pollution.

E. canadensis is native to most of the United States and parts of Canada and was first authentically reported from Great Britain in 1842, having perhaps been introduced on imported timber. Its period of rapid spread started shortly after 1850, when it began to appear in abundance around the Fens, and within a decade it was to be found in most of southern England and parts of the Midlands, extending through nearly all of the English lowlands and into some of southern Scotland by 1880. Subsequent colonisation northwards was more sporadic and largely confined to the east during the closing years of the last century and the beginning of this and in Wales, too, penetration beyond the early stations along the borders was slow and local (Simpson 1984).

E. canadensis seems to have reached the maximum extent of its distribution by the middle of the present century (Simpson 1983, 1986) and remains strongly confined to the lowlands of Britain where more fertile and quieter waters are concentrated in regions of warmer summer climate. Most records occur in places with a mean annual maximum temperature of 24 °C or more (Conolly & Dahl 1970), but within this zone the plant rarely makes vigorous growth unless there is at least modest enrichment of the waters and it performs best in distinctly eutrophic conditions (Haslam 1978, Simpson 1983, 1986). The foliage can actually stand much battering by turbulence (Haslam 1978) but, being poorly anchored or not at all, *E. canadensis* will not persist in waters with a continuously fast flow and favours still or sluggish conditions. In the subdued landscape of the British lowlands, where the majority of the underlying rocks and superficials are readily weathered sedimentaries and open waters accumulate much fine mineral material, these requirements are widely met in both natural and artificial water bodies. Here, in lakes and pools, drainage channels, canals, streams and rivers, stands of this vegetation can penetrate quite deeply if the waters are clear, often to well over 1 m. The substrates are usually silty but *E. canadensis* is intolerant of turbid conditions and does not thrive in shaded waters (Haslam 1978, Simpson 1986). To the north and west, suitable sites of these kinds become scarce, but the community occurs locally in more mesotrophic lakes where there is some sedimentation, usually in deeper waters, over 1 m, where wave disturbance is minimal (Spence 1964).

In the early decades of the expansion of *E. canadensis*, it became clear that invasion and rapid increase in abundance were followed fairly quickly by a decline (Simpson 1984). It might take just three or four seasons for the plant to assume the proportion of a major pest at a site, with such luxuriance being maintained for up to a decade more. Thereafter, though, the population would wane over 7–15 years, leaving just a relict, or disappearing entirely, perhaps returning a few years later. Such a local pattern has been repeated nationally, such that, at the turn of the century, populations were generally declining in the south-east and central England, while invasion was still proceeding vigorously to the north and west. It still seems to be the case that more northerly stands have an abundance not usually seen in the south, although, through the country as a whole, dramatic invasions are now rare events. *E. canadensis* can still colonise new water bodies, however, like flooded gravel pits, and Grose (1957) and Messenger (1971) have both reported the kind of rapid increase in abundance typical of the early invasions.

E. canadensis is a dioecious plant and, with but a single exception, all populations examined have been female (Simpson 1984). Reproduction is thus vegetative with us, occurring by shoot fragmentation, the stems being extremely brittle and the broken portions quickly producing adventitious roots. Expansion within lakes or along canals and rivers is thus fairly simple and perhaps encouraged by periodic disturbance by currents, spates or boat traffic. Dispersal to isolated water bodies is more problematic. Certainly, in the early days of its invasion, deliberate introduction by human agency occurred in some places, as around Cambridge (Marshall 1852), but this never seems to have happened widely and it is possible that waterfowl play an important part. Plants appear to survive for several days out of water, provided conditions remain humid, and fragments of shoots may adhere to feathers or feet (Simpson 1986).

The ability of *E. canadensis* to attain great abundance means that it can become a hazard to water flow through drainage systems and to boat traffic and, despite its reduced presence in many areas nowadays, clearance by chemical or mechanical means is still often necessary. Its power of vegetative spread enables it to recover fairly quickly where fragments are left after cutting or dredg-

ing. Moreover, although under undisturbed conditions *E. canadensis* attains peak biomass in early summer, cut shoots can make luxuriant growth at any time between spring and autumn.

Zonation and succession
The *E. canadensis* community can be found with a variety of other submerged aquatic vegetation in more nutrient-rich, standing or slow-moving waters, although where conditions are especially congenial it may overwhelm these with its luxuriant growth. Open covers of floating-leaved or floating vegetation may occur above the pondweed, and the community can persist in sparse swamp but, where colonising emergents become dense, it is shaded out.

In many lowland ponds, dykes and canals, stands of *E. canadensis* are a major element among the aquatics, sometimes the only submerged vegetation. Duckweed mats are often found in association, of the simpler kind with *Lemna gibba* or *L. minor* or, in purer waters, of the richer *Spirodela-Hydrocharis* type. In a very few places, as among some of the Broadland dykes, the *E. canadensis* community occurs in intimate association with *Hydrocharis-Stratiotes* vegetation. Stands can persist, too, beneath open canopies of the water-lily communities, the *Nymphaeetum albae* and, more commonly in richer waters, the *Nuphar lutea* vegetation, or under the *Potamogeton natans* or *Polygonum amphibium* communities, but wherever these become dense, *E. canadensis* tends to be shaded out. Quite often, in such situations, or beneath overhanging woody vegetation alongside dykes or streams or around pond margins, it is replaced by the more tolerant *Ceratophylletum demersi*.

In other places, there is a more varied submerged aquatic flora in which stands of *E. canadensis* become locally dense among richer assemblages like the *Potamogeton-M. spicatum* community of cleaner, base-rich, eutrophic waters. However, comparing past and present accounts of aquatic vegetation, it seems certain that some of the spread of *E. canadensis* has been at the expense of such communities, their diversity in composition and dominance being reduced as a result. The same trend is visible to a lesser extent in the *Potamogeton-M. alterniflorum* vegetation, though this is a community of more base-poor and less eutrophic waters, many of whose stands occur beyond the range of *E. canadensis*. Both these kinds of pondweed vegetation and related *M. alterniflorum* stands can also tolerate somewhat more fast-moving waters than the *E. canadensis* community and they persist, albeit often in attenuated forms, higher upstream than it can penetrate. The same is true of crowfoot vegetation dominated by *Ranunculus penicillatus* ssp. *pseudofluitans* and *R. fluitans*, though these can be found in patchworks with *E. canadensis* in more slow-moving reaches of streams and rivers, the proportions of the vegetation types often varying from season to season and through any years (Butcher 1933).

E. canadensis is sometimes found growing with its more recently introduced relative E. nuttallii, although great abundance of the one is rarely seen among vigorous plants of the other, and it is possible that there is some kind of competitive relationship between the two. Certainly, there are cases where E. nuttallii has been seen actually to replace E. canadensis over a few years (Briggs 1977, Lund 1979), although the basis of such changes is unclear (Simpson 1984). In polluted waters, neither of these pondweeds thrives and, in the many places where effluents have drained into open waters with the E. canadensis community it has been replaced by P. pectinatus vegetation where any vascular plants at all have survived.

Many of the more fertile lowland waters favoured by E. canadensis are very prone to marginal invasion of swamp plants, and stands are sometimes found among open covers, the Scirpetum lacustris, the Typhetum latifoliae or Glycerio-Sparganion water-margin vegetation, but any more than light shade is inimical to the pondweed.

Distribution
The E. canadensis community is widespread through the lowlands of Britain, though rarely abundant now, particularly in the south of the country.

Affinities
Dense stands of E. canadensis were early recognised as an important component in more fertile aquatic systems (Matthews 1914, Butcher 1933, Tansley 1939) and have also been recorded from the Continent, through most of which the plant is a commonly encountered adventive (Pignatti 1953, Solinska 1963, Passarge 1964). The most obvious location for such vegetation is among the Parvopotamion.

Floristic table A15

Elodea canadensis	V (5–10)
Ceratophyllum demersum	II (2–4)
Lemna minor	II (3–7)
Potamogeton natans	II (1–4)
Myriophyllum spicatum	I (2–3)
Potamogeton pectinatus	I (1–4)
Potamogeton crispus	I (4–6)
Callitriche stagnalis	I (2)
Lemna trisulca	I (2–3)
Glyceria fluitans	I (1–6)
Ranunculus circinatus	I (4)
Potamogeton perfoliatus	I (4)
Ranunculus penicillatus pseudofluitans	I (5)
Callitriche hamulata	I (4)
Polygonum amphibium	I (2)
Spirodela polyrhiza	I (2)
Lemna gibba	I (4)
Potamogeton lucens	I (2)
Nuphar lutea	I (4)
Sparganium emersum	I (5)
Nymphaea alba	I (5)
Number of samples	33
Number of species/sample	2 (1–7)

A17
Ranunculus penicillatus ssp. *pseudofluitans* community

Synonymy
Moderately swift current vegetation   Butcher 1933 *p.p.*

Constant species
Ranunculus penicillatus ssp. *pseudofluitans*.

Physiognomy
This community comprises stands of submerged aquatic vegetation dominated by the crowfoot now known as *Ranunculus penicillatus* ssp. *pseudofluitans*, a taxon previously designated as *R. penicillatus* ssp. *calcareus* (Butcher 1960, Cook 1966, Holmes 1980, Webster 1988). It is a perennial plant with shoots up to 3 m long, growing in clumps and trailing downstream in the flowing water that provides the usual habitat. Maximum cover is attained early in the season, with luxuriant growth of the fine tasselled leaves occurring where conditions are congenial, and clumps can be very numerous such that whole streams of considerable size become choked with this vegetation. The var. *vertumnus*, sometimes separated from a var. *pseudofluitans* (Holmes 1980, Webster 1988), was not distinguished in the sampling.

The community is frequently found in close association with other crowfoot vegetation, like that dominated by *R. aquatilis* and *R. peltatus*, and various other aquatic assemblages, but among denser stands associates are usually few and sparse. *Callitriche stagnalis* and *Potamogeton pectinatus* are occasionally seen, and there are sometimes small patches of *Lemna minor* caught among floating shoots where the flow is slacker. Then, there are quite often some trailing or emergent shoots of plants such as *Nasturtium officinale*, *Veronica beccabunga* and *Berula erecta*, all of which tend to become more luxuriant rather later in the season, as the abundance of the crowfoot is fading.

Habitat
The community is confined to base-rich but generally only moderately fertile waters, of moderate to quite fast flow and with sandy to gravelly beds, mostly in limestone catchments in lowland England and Wales, and especially towards the south.

As diagnosed by Holmes (1979, 1980) and Webster (1988), *R. penicillatus* ssp. *pseudofluitans* is centred in southern England, particularly in waters on the Chalk and Oolite, with some occurrences also on London and Oxford Clays. Further west and north, it can be found on Carboniferous Limestone, Devonian and Silurian rocks, which provide most of the substrates in Wales, but it scarcely penetrates into Scotland, being found in just a very few localities on Old Red Sandstone. Such rocks produce the necessary lime-rich drainage waters for this kind of crowfoot, which is usually found where the pH is between 7 and 9 and the alkalinity 100–300 mg l^{-1} calcium carbonate, somewhat lower in less calcareous river systems like the Tweed and the Usk (Webster 1988). In such waters, nutrient status under natural conditions is generally only moderate and the preference of *R. penicillatus* ssp. *pseudofluitans* seems to be for mesotrophic to fairly eutrophic situations (Haslam 1978, Newbold & Palmer 1979). Where there is a switch from limestone to clays in the catchment, for example, this vegetation is usually lost before the latter produce half of the drainage waters (Haslam 1978), and cultural eutrophication is inimical to the vigour of the community.

Sands and gravels are the preferred substrates and, in such coarser material, the plants are quite difficult to erode despite being shallow-rooted. Although occasional in sluggish conditions, even in dykes and pools, *R. penicillatus* ssp. *pseudofluitans* thus often occurs in waters of quite fast flow, extending into the upper reaches of streams provided these are not too prone to very fierce spates or do not dry up in summer. This kind of vegetation is concentrated in depths of less than 1 m, although bigger volumes of water seem to be preferred (Haslam 1978).

Where conditions are favourable, the plants can quickly spread (Holmes & Whitton 1977a, b) and become established over extensive stretches of river, when the resistance to flow of luxuriant growth often results in back-up of the waters and flooding. Considerable effort is therefore needed in some drainage systems to control this vegetation by chemical or mechanical means (Dawson 1978, Westlake & Dawson 1982, Webster 1988) and great attention has been given to the impact of abundant stands of the community on nutrient and energy cycling in rivers (Edwards & Owens 1960, Owens & Edwards 1961, 1962, Westlake 1975, Holmes *et al.* 1972, Casey & Downing 1976). Since regrowth after early cutting can be very vigorous, care needs to be taken not to cut unnecessarily at the beginning of summer if this simply provokes a flood hazard later in the season (Haslam 1978).

Zonation and succession
R. penicillatus ssp. *pseudofluitans* vegetation is sometimes found dominating whole stretches of streams and rivers to the virtual exclusion of other aquatics, but often it occurs with other assemblages of submerged and floating-leaved plants in patterns related to the speed of flow and trophic state of the waters. Zonations along and across Chalk streams are especially distinctive with seasonality of flow becoming important in the upper reaches.

Where this kind of vegetation extends into slacker waters it shows some overlap in its occurrence with *Elodea canadensis* stands and the *Ceratophylletum demersi*, the former like the crowfoot tending to be most prominent in early summer, the latter becoming more abundant later in the season. There can also be considerable differences in the proportions of the communities from season to season (Butcher 1933). With increase in the trophic state of the waters and a shift to finer sediments, as happens in moving downstream, it is vegetation of this kind that eventually replaces the *R. penicillatus* ssp. *pseudofluitans*, or there may be a switch to richer *Potamogeton-M. spicatum* vegetation, a pattern well seen with the move to clay substrates in the lower reaches of some Chalk streams. The *Nuphar lutea* community can also become prominent over crowfoot stands as such quieter, richer waters are reached.

Callitriche stagnalis stands are common, too, in shallower or sheltered places in these stretches but it is in the higher, faster reaches that this vegetation becomes really abundant in Chalk streams, forming a zone along the water margins either side of a central strip often dominated by *R. penicillatus* ssp. *pseudofluitans*, but with locally prominent *Ranunculetum aquatilis* and *Ranunculetum peltati*. Patches of herbs like *Berula erecta* and *Apium nodiflorum* can be abundant among the submerged crowfoots, particularly as these lose their vigour later in the season, and they often thicken up in shallower margins, together with *Nasturtium officinale*, to form a semi-emergent band of Glycerio-Sparganion vegetation, with occasional clumps of *Sparganietum erecti* swamp. Higher still, where summer flow becomes sporadic or non-existent, *R. pencillatus* ssp. *pseudofluitans* tends to yield dominance in the central zone to the *Ranunculetum aquatilis* and *Ranunculetum peltati* with patches of *Callitriche* vegetation, and the Glycerio-Sparganion herbs may extend in irregular fashion over much of the stream bed.

R. penicillatus ssp. *pseudofluitans* is only rarely represented in the aquatic assemblages of more base-poor streams and rivers where *Callitriche* and *Myriophyllum alterniflorum* vegetation are the major elements in faster reaches, but ssp. *penicillatus* (Webster 1988) is locally found there, and further sampling is needed through western England and Wales to establish the character of its stands (Newbold & Palmer 1979, Holmes 1983).

Distribution
The community is centred in southern England, extending northwards in more scattered localities into south Scotland and westwards into Wales.

Affinities
The confusing taxonomic history of the various crowfoots makes it difficult to trace the role of *R. penicillatus* ssp. *pseudofluitans* in earlier accounts of river vegetation, but the plant was generally included in broadly defined assemblages of moving water (e.g. Butcher 1933, Tansley 1939). Exactly comparable associations have not been characterised from the Continent, but the community belongs to the Callitricho-Batrachion (or Ranunculion fluitantis in Ellenberg 1978).

Floristic table A17

Ranunculus penicillatus pseudofluitans	V (3–10)
Lemna minor	II (3–7)
Callitriche stagnalis	II (2–6)
Nasturtium officinale	II (3–4)
Veronica beccabunga	II (1–3)
Berula erecta	II (1–4)
Potamogeton pectinatus	I (3)
Polygonum amphibium	I (3)
Potamogeton crispus	I (4)
Polygonum hydropiper	I (2)
Elodea canadensis	I (1)
Number of samples	13
Number of species/sample	2 (1–5)

A18
Ranunculus fluitans community
Ranunculetum fluitantis   Allorge 1922

Synonymy

Moderate-moderately swift current vegetation Butcher 1933.

Constant species

Ranunculus fluitans.

Physiognomy

The *Ranunculetum fluitantis* comprises stands of submerged vegetation dominated by clumps of *Ranunculus fluitans*, sometimes numerous and close-set, in other cases few and sparse. It is generally a perennial plant, with individuals often long-lived though highly plastic in their morphology with just a few shoots, fairly short, in less congenial situations, but growing much more bushy and very long, up to 6 m, where conditions are more favourable, with the fine capillary foliage trailing downstream. Rarely, it can be found in an annual terrestrial form, with much-condensed shoots growing on moist ground (Cook 1966, Holmes 1979, Rich & Rich 1988). It is winter-green, with quite large populations of shoots persisting from one year to the next, but attains its maximum size early in the summer (Haslam 1978).

Few other plants occur with any frequency or abundance in denser stands, but there is sometimes a little *Myriophyllum spicatum*, *M. alterniflorum* or *Potamogeton perfoliatus* with patches of the mosses *Fontinalis antipyretica* or *F. squamosa* growing on submerged stones. *Elodea canadensis*, *Lemna minor* or *L. gibba* may occur in slacker waters, and there can be shoots of plants like *Glyceria fluitans* and *Mentha aquatica* trailing in from the margins.

Habitat

The *Ranunculetum fluitantis* is most characteristic of bigger moving waters, often with quite swift flow, and stable, stony beds, usually only moderately fertile and not very base-rich. It is virtually confined to England where it occurs most commonly in faster lowland streams and wide but not too spatey, rivers in the upland fringes.

R. fluitans needs considerable water movement to maintain good growth, and deeper channels, often of 1 m or more, are favoured (Haslam 1978). It will occasionally colonise sluggish waters in larger dykes and drains, though it rarely flowers in such situations (Holmes 1979), and it is very easily eroded from finer, soft substrates. It is much more frequent on consolidated beds, with gravel or pebbles, sometimes larger stones or boulders, and here it can gain a much firmer hold. The long, flexuose shoots, growing in streamlined clumps, offer little resistance to turbulence, and *R. fluitans* can extend into quite fast-flowing waters. It will persist even in torrents as sparse shoots but will not tolerate very spatey conditions, so in hill steams tends to be best developed in wider places where the bed is more stable. Towards the lowlands, the favoured, firmer substrates are usually concentrated in medium-width stretches of rivers (Haslam 1978).

Even in southern Britain, though, to which pure *R. fluitans* seems to be restricted (Holmes 1979), this vegetation is not usually found in more calcareous or eutrophic waters, being most typical of mesotrophic to quite oligotrophic and sometimes fairly base-poor conditions, extending into streams on impoverished and acidic rocks around the Welsh Marches, the Pennines and the Lake District (Haslam 1978). The faster flow and stony substrates help maintain the favourable environment, and eutrophication of rivers by draining in of enriched ground waters or effluents is inimical to *R. fluitans*.

Zonation and succession

The *Ranunculetum fluitantis* can sometimes monopolise the vegetated stretches in faster, stony-bedded rivers but it is often found in mosaics with other aquatic communities where there is variation in turbulence and depth of the waters and the character of the substrate, and gives way to other assemblages with a shift in these factors downstream and up.

Where the flow becomes a little slacker and the bed finer, this kind of vegetation can be found with stands of *Elodea canadensis* and, where the waters are less base-poor, the *Potamogeton-M. spicatum* community, and these may replace the *Ranunculetum fluitantis* where lower river catchments are dominated by shales or clays. In highly eutrophic steams, *Potamogeton pectinatus* stands can take over from the community. Floating-leaved stands of *Potamogeton natans* or *Polygonum amphibium* may occur and in the shallows there can be a zone of Glycerio-Sparganion vegetation or patches of the *Phalaridetum*.

These often persist along the margins of moving waters upstream of the major *Ranunculetum fluitantis* stretches where the submerged vegetation tends to comprise mixtures of *Callitriche stagnalis* stands, sometimes with the *Ranunculetum aquatilis* or, in more base-poor streams, the *Potamogeton-M. alterniflorum* or *M. alterniflorum* communities.

Distribution

The community is concentrated in northern and central England, with more local sites further south, along the Welsh Marches and just beyond the Scottish border.

Affinities

R. fluitans has generally been recorded in accounts of British vegetation as part of more broadly-defined assemblages of moving waters (e.g. Butcher 1933, Tansley 1939) but our stands are clearly comparable with the *Ranunculetum fluitantis* described from northern France (Allorge 1921–2), The Netherlands (Westhoff & den Held 1969) and Germany (Oberdorfer 1977) and placed in the Callitricho-Batrachion (or Ranunculion fluitantis as Ellenberg (1978) has it).

Floristic table A18

Ranunculus fluitans	V (4–9)
Myriophyllum spicatum	II (4–5)
Glyceria fluitans	I (1)
Elodea canadensis	I (4)
Lemna minor	I (5)
Fontinalis antipyretica	I (5)
Potamogeton perfoliatus	I (6)
Fontinalis squamosa	I (6)
Myriophyllum alterniflorum	I (1)
Mentha aquatica	I (2)
Phalaris arundinacea	I (2)
Lemna gibba	I (1)
Number of samples	8
Number of species/sample	2 (1-4)

A19
Ranunculus aquatilis community
Ranunculetum aquatilis Géhu 1961

Synonymy
Vegetation of nearly stagnant waters Tansley 1911
p.p.

Constant species
Ranunculus aquatilis.

Physiognomy
The *Ranunculetum aquatilis* is dominated by clumps or patches of *Ranunculus aquatilis*, a very variable annual or perennial crowfoot, able to grow submerged, when the shoots are spreading to erect, as a delicate floating plant or terrestrially on moist ground, when it is loosely cespitose (Haslam 1978, Rich & Rich 1988). Cover is very variable and many stands are small, but in suitable conditions this species can make luxuriant growth, becoming most abundant in early summer.

The vegetation often grows in intimate association with various other aquatics and amphibious plants, including stands of *Ranunculus peltatus*, which is morphologically very similar to and readily confused with *R. aquatilis* (Holmes 1979, Rich & Rich 1988). But the most frequent companions in denser stands are various stoneworts, particularly *Callitriche stagnalis* and *C. obtusangula*, patches of which can become prominent with the crowfoot early in the season, and Glycerio-Sparganion herbs, which increase their cover later. Commonest among these latter are smaller *Glyceria* spp., especially *G. fluitans*, and *Apium nodiflorum*, but *Nasturtium officinale*, *Veronica beccabunga* and *Berula erecta* can also be found, and each of these can grow in vigorous patches among the submerged plants or as floating or emergent shoots. *Myriophyllum spicatum*, *Ceratophyllum demersum* and *Elodea canadensis* are sometimes seen, and there can be floating-leaved mats of *Potamogeton natans* with patches of duckweed thalli caught among the vegetation on the water surface or wet mud.

Habitat
The *Ranunculetum aquatilis* is typically found in and around the margins of mesotrophic to fairly nutrient-rich waters, sometimes quite fast-moving, in other cases standing or sluggish. It probably occurs through much of southern Britain outside the highland zone, extending into streams around the upland fringes, but becoming much more common in streams, dykes, canals and pools in the lowlands. It tolerates periodic or seasonal drying and will colonise disturbed or ephemeral water-margin habitats.

The difficulty of identifying *R. aquatilis* with certainty and in particular the problem of separation from *R. peltatus* (Holmes 1979, Rich & Rich 1988), makes it hard to define precisely the geographical or environmental limits of this vegetation. It seems to occur through most of the lowlands and the upland fringes, and is concentrated in shallower waters, often much less than 1 m deep and sometimes fluctuating, with fertility that avoids the extremes of impoverishment and marked eutrophication. *R. aquatilis* has tough anchoring rhizomes, becomes well rooted in sandy or gravelly substrates and has fine-leaved submerged shoots that are very resistant to turbulence, so it can extend into quite fast-flowing waters (Haslam 1978).

In the foothills of the north and west, then, the *Ranunculetum aquatilis* can be seen towards the upper reaches of streams, and there the substrates are generally resistant rocks with drainage waters that are more acidic and nutrient-poor. The community becomes much more common, however, in the south and east of the country, where the waters tend to be more fertile, and generally more base-rich. It is quite well represented, for example, in the mesotrophic stretches of Chalk streams where winter flow can be fairly rapid, but it occurs widely, too, in the more sluggish waters of canals, dykes and pools. Its frequent presence in farm ponds has led to the suggestion that *R. aquatilis* may prefer more eutrophic habitats than *R. peltatus* (Rich & Rich 1988), but the community disappears from highly enriched and polluted waters.

The tolerance that *R. aquatilis* shows for periodic or seasonal drying of the ground gives it an important advantage in fluctuating waters. This kind of vegetation will stand the summer droughting in the upper reaches of Chalk streams, for example, and persist around the margins of ponds that dry up from time to time, provided it is not shaded by bulkier opportunist invaders. The disturbance that grazing stock provide around farm ponds is probably important in maintaining more open moist ground on which this vegetation can prosper.

Zonation and succession

The *Ranunculetum aquatilis* can be found with various other kinds of submerged aquatic vegetation, and sometimes with floating-leaved communities, the mosaics and zonations being influenced by the depth and speed of the waters and the character of the substrates. Towards water margins, transitions to Glycerio-Sparganion vegetation or certain types of swamp are usual, and taller emergents eventually overwhelm the community where they are invading shallower waters. Often, however, turbulence or fluctuation of the waters, or disturbance of the margins, helps maintain the *Ranunculetum aquatilis* as a more or less permanent feature.

In faster-moving shallows, the community may be the major element among the submerged vegetation of coarser sands and gravelly beds, or occur with patches of the *Callitriche stagnalis* vegetation. In more base-poor streams, such as are found through the upland fringes of the north and west, the *Myriophyllum alterniflorum* community can also figure, these two kinds of vegetation generally extending much further into the torrential and spatey reaches upstream. In deeper waters with stable, stony beds, the *Ranunculetum aquatilis* can give way to the *Ranunculetum fluitantis*, and this vegetation may replace it entirely in wider stretches of river downstream, where the substrate consists of consolidated pebbles and boulders. Slacker reaches, with some deposition of finer mineral material, can have stands of the *Potamogeton-M. alterniflorum* community and floating-leaved *Nuphar lutea* or *Potamogeton natans* vegetation. Along the water margins, there are often fragments of Glycerio-Sparganion assemblages or *Phalaridetum* swamp.

In more calcareous, swift-flowing waters, notably in Chalk streams, the *Ranunculetum aquatilis* is not usually so abundant as the *Ranunculetum peltati* (Haslam 1978), but both these vegetation types comprise an important element in the stream beds of the upper reaches, often mixed with *Callitriche stagnalis* stands, and generally

bordered or interrupted by clumps of Glycerio-Sparganion communities. The ability of the two crowfoot assemblages to withstand the late-summer drying-up of these streams gives them an important competitive advantage against *Ranunculus penicillatus* ssp. *pseudofluitans*, though this becomes increasingly important with the move downstream into reaches that have perennial flow. With the shift into the more eutrophic stretches of these streams, where there is often much clay amongst the substrate, and where vegetation like the *Potamogeton-M. spicatum* community can become important, these crowfoot communities generally disappear.

In the shallows of sluggish and standing waters of quite high fertility, the *Ranunculetum aquatilis* can maintain a presence, though it sometimes assumes a frail, floating form and often makes but a small contribution among luxuriant stands of submerged vegetation like the *Elodea canadensis*, *Ceratophyllum demersum* and *Potamogeton pectinatus* communities. However, fluctuation in such waters or disturbance of the margins can favour its persistence and, around the margins of such pools and dykes, it often occurs in patchy mosaics with *Callitriche stagnalis* vegetation, sparse duckweed mats, Glycerio-Sparganion herbs and *Sparganietum erecti* swamp. In these situations, succession is repeatedly set back, but where more slow-moving waters are prone to invasion by emergents, the *Ranunculetum aquatilis* is quickly shaded out in dense covers.

Distribution

Careful diagnosis of *R. aquatilis* is needed to establish the exact distribution of this vegetation, but it is probably widespread through the lowlands and upland fringes of Britain.

Affinities

Early descriptive accounts of British aquatics included stands of *R. aquatilis* among more broadly defined assemblages (e.g. Tansley 1911) and in phytosociological schemes it has sometimes been included with *R. peltatus* vegetation in a *Ranuculetum peltati* (Segal 1967, Westhoff & den Held 1969). Other Continental workers have maintained distinct associations for the two species (Sauer 1937, Géhu 1961). Whichever solution is adopted, this kind of vegetation clearly belongs with other crowfoot assemblages in the Callitricho-Batrachion (or Ranunculion fluitantis of Ellenberg 1978).

Floristic table A19

Ranunculus aquatilis	V (2–8)
Glyceria fluitans	III (2–5)
Apium nodiflorum	II (1–2)
Myriophyllum spicatum	II (5–6)
Fontinalis antipyretica	II (1–4)
Potamogeton natans	II (1–5)
Callitriche stagnalis	II (3)
Callitriche obtusangula	II (4–7)
Glyceria × *pedicellata*	II (5)
Callitriche hamulata	I (4)
Lemna gibba	I (4)
Myriophyllum alterniflorum	I (4)
Ceratophyllum demersum	I (2)
Ranunculus fluitans	I (2)
Elodea canadensis	I (3)
Potamogeton pectinatus	I (5)
Lemna trisulca	I (2)
Polygonum amphibium	I (2)
Number of samples	8
Number of species/sample	6 (3–8)

A20

Ranunculus peltatus community
Ranunculetum peltati   Sauer 1947

Synonymy

Vegetation of nearly stagnant waters   Tansley 1911 *p.p.*

Constant species

Ranunculus peltatus.

Physiognomy

The *Ranunculetum peltati* is dominated by clumps or patches of *Ranunculus peltatus*, the most variable of the British crowfoots. It can grow as an annual or perennial, submerged in a spreading or erect form, as a frail, floating plant or occur terrestrially, when it is loosely tufted (Haslam 1978, Rich & Rich 1988). Many stands are small and cover can be very sparse, but luxuriant plants develop in congenial situations with maximum growth in early summer.

The *Ranunculetum peltati* is frequently found in close association with other kinds of aquatic and marginal vegetation, including patches of the *Ranunculetum aquatilis*, and it can be very difficult to distinguish the two dominants from one another (Holmes 1979, Rich & Rich 1988). In denser stands, however, associates are few and usually not very abundant and such plants as do occur are often amphibious or water-margin herbs. *Lemna minor* and *L. gibba* are often found on the surface of the water or moist mud, with *L. trisulca* occasionally caught among the submerged shoots, and there can be patches of *Callitriche* spp., *C. platycarpa* in the stands sampled. *Potamogeton natans* and *Polygonum amphibium* are occasionally seen and sometimes there are plants of other *Ranunculus* spp., such as *R. sceleratus*, *R. trichophyllus* and *R. circinatus*. Later in the season, Glycerio-Sparganion plants often become prominent with patches of *Nasturtium officinale*, *Glyceria fluitans*, *G. plicata*, *Apium nodiflorum* and *Berula erecta*.

Habitat

The *Ranunculetum peltati* is characteristic of the shallows and margins of mesotrophic to quite nutrient-rich waters, occasionally fairly fast-flowing, though usually sluggish or still. It appears to occur through most of southern Britain outside the highland areas, but is especially frequent in the English lowlands. It withstands seasonal or periodic drying out of the habitat and this enables it to persist in fluctuating or ephemeral water bodies.

R. peltatus can be hard to distinguish from *R. aquatilis* (Holmes 1979, Rich & Rich 1988) and the two taxa are clearly similar in their geographical distribution and ecological preferences, though it has sometimes rather puzzlingly been reported that they do not in fact occur together (Cook 1966). Like the *Ranunculetum aquatilis*, this vegetation is commonly found growing submerged, and it may penetrate a little deeper (Holmes 1979), although most stands seem to occur in less than 1 m of water. It is perhaps also less demanding of fertile conditions, so it is quite frequent around the margins of lakes and pools through the upland fringes of the north and west of the country, where there is but modest amelioration of nutrient impoverishment. Here, it will extend on to coarser mineral substrates, such as sands and gravels, and it can be difficult to erode from such materials where the tough rhizomes and root wefts gain a strong hold, so it can tolerate some gentle surge along shores and is sometimes found in fairly swift-moving streams (Haslam 1978).

In the south-west, too, it is an important community of fairly fast-flowing waters of a calcareous nature, notably in the upper reaches of Chalk streams, and is perhaps more characteristic of base-rich habitats than the *Ranunculetum aquatilis* (Holmes 1979). Here, again fertility is only moderate, but the community also occurs widely in sluggish lowland waters which are more eutrophic, as in canals, dykes and pools, where the substrates are more silty. Its tolerance of periodic drying gives this vegetation an important advantage in Chalk streams which cease to flow in summer and around the margins of fluctuating or temporary water bodies. Disturbance of such ground helps keep the habitat sufficiently open for *R. peltatus* to thrive.

A22
Littorella uniflora-Lobelia dortmanna community

Synonymy
Littorella-Lobelia associes Pearsall 1918; *Eleocharitetum acicularis* Koch 1926 *p.p.*; *Isoeto-Lobelietum* (Koch 1926) Tx. 1937 *p.p.*; *Eleocharitetum multicaulis* Tüxen 1937 *p.p.*; *Juncus fluitans-Lobelia dortmanna* and *Lobelia dortmanna* associations Spence 1964; *Littorella uniflora-Lobelia dortmanna* Association Birks 1973; *Eriocaulo-Lobelietum* Br.-Bl. & Tx. 1952 *sensu* Birse 1984.

Constant species
Littorella uniflora, *Lobelia dortmanna*.

Rare species
Elatine hexandra, *Eleocharis acicularis*, *Eriocaulon septangulare*, *Isoetes setacea*, *Subularia aquatica*.

Physiognomy
The *Littorella uniflora-Lobelia dortmanna* community comprises open or closed swards of submerged or temporarily emergent vegetation, usually less than 10 cm tall, dominated by gregarious rosette plants with linear or subulate leaves. The commonest of these is *Littorella uniflora*, particularly obvious in younger stands and in shallower, wave-churned waters, sometimes sparsely scattered over stony ground, but often dominant in fine densely-packed lawns. *Lobelia dortmanna* is the only other constant and it, too, can grow in quite luxuriant profusion here, frequently exceeding *Littorella* in more sheltered, somewhat deeper waters, and revealing its presence in mid-summer with its emergent racemes of attractive lilac flowers. It is, though, not so widely distributed geographically as *Littorella* and is altogether absent from those fragmentary stands of the community found in southern England.

Also quite common throughout the range of this vegetation is *Juncus bulbosus*, often occurring in its free-floating form and sometimes forming thick tangles of slender shoots, but no other associate is frequent overall. Thus, there are occasionally some emergent shoots of *Carex rostrata*, *Equisetum fluviatile* and, in shallower waters, *Eleocharis palustris* and the more local *E. multicaulis*, but increased frequency and cover of these plants usually marks a shift to swamp vegetation. Likewise, there can be some *Myriophyllum alterniflorum* but this is strongly associated with stands in deeper and less turbulent waters, where the occasional presence of *Isoetes lacustris* or, more locally, *I. setacea* marks a transition to the zone in which these quillworts become dominant.

Other plants which occur sparsely through the community are *Scirpus fluitans*, which is often difficult to see when floating among masses of *J. bulbosus*, and *Baldellia ranunculoides*, the emergent leaves of which often get abraded by wave action. This kind of vegetation also provides a locus for the rare annual *Elatine hexandra*, for *Eleocharis acicularis* in Scotland and, in northern and western Britain, for *Subularia aquatica*. This is another very local therophyte, occasionally overwintering, that is readily overlooked among young *Littorella* but able to attain some abundance on open but stable substrates (Woodhead 1951b). A further rarity, *Eriocaulon septangulare*, can also be found in this community in some of its very few Scottish localities. It is a widespread plant in this sort of vegetation in western Ireland (Braun-Blanquet & Tüxen 1952), but occurs with us only in Inverness and on Coll and Skye (Perring & Farrell 1977), and even then not always among *Littorella-Lobelia* swards. On Skye, for example, where dense mats occur most commonly, it is more usually seen among stands of the *Caricetum rostratae* (Birks 1973).

Sub-communities

Littorella uniflora **sub-community:** *Eleocharitetum multicaulis* Tüxen 1937 *p.p.*; open *Littorella-Lobelia* sociation Spence 1964; *Littorella uniflora-Lobelia dortmanna* Association Birks 1973; *Isoeto-Lobelietum* (Koch 1926) Tx. 1937, inops and *eleocharetosum* Schoof-van Pelt 1973; *Isoeto-Lobelietum* (Koch 1926)

Tx. 1937 *sensu* Birse 1984; *Eriocaulo-Lobelietum* Br.-Bl. & Tx. 1952 *sensu* Birse 1984. *Littorella* is the commonest species here, sometimes virtually the only plant in very open stands, though usually quite abundant, and generally exceeding *L. dortmanna*, which can be absent altogether. Locally, *Isoetes lacustris* is prominent but quillworts are more characteristic of the other sub-community, together with *Myriophyllum alterniflorum* which is very scarce here.

There is frequently some *Juncus bulbosus* and, among emergent plants, occasional *Carex rostrata* and *Equisetum fluviatile* are quite often accompanied here by *Eleocharis palustris* and, more locally, by *E. multicaulis*, these last two sometimes attaining high cover. In shallower waters, *Ranunculus flammula* may be prominent, together with plants like *Juncus articulatus* and *Carex nigra*. It is in this kind of *Littorella-Lobelia* vegetation that *E. septangulare* has been recorded, sometimes as a local dominant.

***Myriophyllum alterniflorum* sub-community:** *Eleocharitetum acicularis* Koch 1926 *p.p.*; *Lobelia-Littorella* and *Littorella-Juncus* sociations Spence 1964; *Isoeto-Lobelietum* (Koch 1926) Tx. 1937 *sensu* Schoof van Pelt 1973 *p.p.*, *sensu* Birse 1980. Although stands can be found here in which either *Littorella* or *L. dortmanna* is absent, the two plants are equally common overall, both can be plentiful and each may dominate. *J. bulbosus* is frequent and there is characteristically some *M. alterniflorum*, occasionally in abundance. *I. lacustris* and the much more local *I. setacea* are also weakly preferential to this sub-community and each can have high cover, particularly in transitions to quillwort swards.

Occasionally there is some emergent *C. rostrata* and *E. fluviatile* but *Eleocharis palustris* and *E. multicaulis* are only infrequently found. However, *E. acicularis* has been recorded in this vegetation in some of its Scottish localities and *Subularia aquatica* is sometimes seen here.

Habitat

The *Littorella-Lobelia* community is characteristic of the barren, stony shallows of clear and infertile standing waters. It is strongly concentrated in the north and west of Britain, where it is a widespread and common feature of lakes and pools, often extending around more exposed shores where there is some wave disturbance, but tolerating only brief exposure where water-levels fall in summer.

Through Europe as a whole, both *Littorella* and *L. dortmanna* have northerly ranges (Matthews 1955) and, in this country, they are largely confined to the north and west. This is especially true of *L. dortmanna*, which requires a cool but equable, northern oceanic climate, such that in the more frigid waters of higher altitudes

and towards the warmer south of the country, stands of the community have a more fragmentary composition in which mixtures of *Littorella* and *J. bulbosus* tend to be the most obvious element. To some extent, though, the concentration of this kind of vegetation in the north and west is a reflection of the distribution of suitable substrates and waters, because clear, oligotrophic lakes are much more common in the glaciated catchments of acidic, resistant rocks that prevail there. Throughout its range, then, this is a community of infertile waters, with conductivities generally less than 200 μmho, pH usually from 5.5 to 7 and alkalinities below 25 mg l^{-1} calcium carbonate (Woodhead 1951*a*, Spence 1964, Palmer 1992, Palmer *et al.* 1992). Locally, such conditions are met in the south and east of Britain in lowland pools over base-poor, sandy substrates, as in Hatchet Pond in the New Forest and Little Sea Mere in Dorset (Ratcliffe 1977). But it is in open waters through the uplands, particularly from Snowdonia northwards, that this kind of vegetation becomes really frequent, sometimes occupying most of the bed of shallow, stony pools, but generally forming a fringe around the edge of larger lakes. Sands and gravels are the preferred substrates, sometimes with a few centimetres of organic detritus or peaty mud, but typically without any fine mineral material. Exposure to wave action in the shallows hinders the deposition of such silt as may be washed into the lakes, helping to maintain infertility.

Of the two constants, *Littorella* seems to be the earlier coloniser of bare, coarse and sometimes rather shifting substrates (Pearsall 1921) and to be the more tolerant of the periodic exposure of lake margins that occurs in drier summers. Indeed, it only flowers when its scapes, which are shorter than the foliage, are emergent. Generally, however, after initial establishment, it spreads by vegetative means, producing abundant far-creeping stolons that readily root and put up rosettes of leaves at the nodes. In this way, *Littorella* can quickly form a dense lawn and retain much of its vigour from season to season.

L. dortmanna, it seems, enters subsequently, each plant establishing from seed and needing some consolidation of a finer mineral bed, perhaps with some organic detritus caught on the surface, before it can establish firmly (Pearsall 1921, Woodhead 1951*a*). It can colonise quite shallow waters, just 10 cm or so deep, but it is generally intolerant of desiccation, being reduced in vigour when water levels are low in winter and spring, and readily killed if exposed in summer (Sylven 1903). Being never very strongly rooted, it is also rather more susceptible than is *Littorella* to the wave-turbulence of exposed shallows and, lacking any means of vegetative expansion, damage to the sward is made good only slowly. It certainly grows most luxuriantly in more sheltered and somewhat deeper waters and, unlike *Lit-*

torella, can put up its first flowers above the surface from considerable depths.

The shift in relative importance of the two constants in these contrasting situations is reflected among the associates and contributes to the definition of the two sub-communities. The *Littorella* type includes stands of the community outside the geographical range of *L. dortmanna*, but generally it is characteristic of shallower and more exposed waters, where the preferentials are those emergents able to form a sparse cover in the stony and wave-torn conditions. The *M. alterniflorum* sub-community can extend into shallows, but only where these are more sheltered, and it is usually found at somewhat greater depths, perhaps just a few decimetres below the *Littorella* type, sometimes down to 1 m or more, but where turbulence is reduced enough to allow some consolidation of the mineral base and the accumulation of a primitive organic layer. At such depths, it is able to extend on to the more exposed sides of lakes, where the shallows are too wave-churned to support even the *Littorella* sub-community. It is the stiller conditions here that favour the growth of *M. alterniflorum*, with the appearance of the quillworts reflecting the more frequent occurrence of stable substrates at some depth.

Zonation and succession
The *Littorella-Lobelia* community is a common element in the very characteristic zonations that are found in less fertile stretches of standing waters, the proportions and dispositions of the various vegetation types reflecting differences in water depth, turbulence and the character of the substrates. In more exposed and oligotrophic situations, the community is a more or less permanent feature in successions that are held in check by the harsh environmental conditions but, where accumulation of silt or organic detritus is able to proceed, this kind of vegetation is replaced by other aquatic communities or swamp.

Quite often, both sub-communities of the *Littorella-Lobelia* vegetation occur together, the *Littorella* type occupying more shallow or turbulent waters, the *M. alterniflorum* sub-community replacing it with increasing depth or shelter, moving away from the shore or shifting round to less exposed aspects. Where more or less stony substrates run on into even deeper water in upland lakes, where neither *Littorella* nor *L. dortmanna* can maintain a vigorous presence, the community often gives way to *Isoetes* vegetation, the quillworts assuming dominance and extending as sometimes pure swards to several metres depth beneath the lower limit of the *Littorella-Lobelia* community (Gay 1863, Pearsall 1921, Spence 1964).

In lakes which are not quite so infertile, however, or where there are local areas of shelter and deposition of finer mineral material within the waters, the *Littorella-*

Lobelia swards can give way to other communities of submerged aquatics able to take advantage of the enriched conditions (Figure 8). Where there are banks of silt along lee shores, for example, or in somewhat deeper waters beyond the wave-disturbed zone or where currents from incoming steams drop suspended, this kind of vegetation is often replaced by the *Potamogeton-M. alterniflorum* community. There, both *Littorella* and *M. alterniflorum* remain very frequent, but dominance commonly passes to pondweeds such as *Potamogeton perfoliatus, P. gramineus, P. berchtoldii, P. obtusifolius* or *P. pusillus*. Dense stands of *M. alterniflorum* can punctuate such patterns, often where there is a slight local increase in turbulence over sandier substrates, and there can be floating-leaved *Nymphaeetum albae* and *Potamogeton natans* vegetation with the shift to more sheltered and deeper waters. The zonations described from Cumbrian lakes by Pearsall (1918, 1921) show such variations clearly.

In more base-rich, calcareous waters, the *Littorella-Lobelia* community often maintains a presence in the more wave-torn shallows, giving way in analagous zonations to those described above to the *Potamogeton-M. spicatum* community, sometimes with *P. pectinatus* stands and floating-leaved *Nuphar lutea* vegetation. Increased eutrophication of lakes with these kinds of patterns has seen an increase in recent years in the amounts of *Elodea canadensis* and *E. nuttallii* among such assemblages.

Towards the opposite extreme, where waters become more peaty and dystrophic, as where lakes occur among or receive drainage waters from stretches of blanket mire, a common feature in Shetland and along parts of the north-west Scottish seaboard, the *Littorella-Lobelia* community can give way in sheltered shallows to the *Juncus bulbosus* vegetation. Sometimes this can be found suspended over sparse *Littorella-Lobelia* swards in peat-bound embayments with stony beds, but it is *J. bulbosus* and *M. alterniflorum* which provide the strongest continuity between the vegetation types, with species such as *Potamogeton polygonifolius, Utricularia vulgaris* and *Sphagnum auriculatum* becoming more abundant where there is a real shift in the character of the habitat (Spence 1964). In the striking machair lochs on South Uist and at a few localities on the nearby Scottish mainland, the *Littorella-Lobelia* community can be found in the same water bodies as both dystrophic vegetation of this kind and basiphilous pondweed assemblages on calcareous shell-sand (Ratcliffe 1977).

Where exposure restricts the accumulation of silt or organic detritus, the extent of *Littorella-Lobelia* stands can remain unchanged for decades (Spence 1964), but more sheltered and shallow waters are always prone to invasion by emergents. *Eleocharis palustris* and, more locally, *E. multicaulis* seem especially common as col-

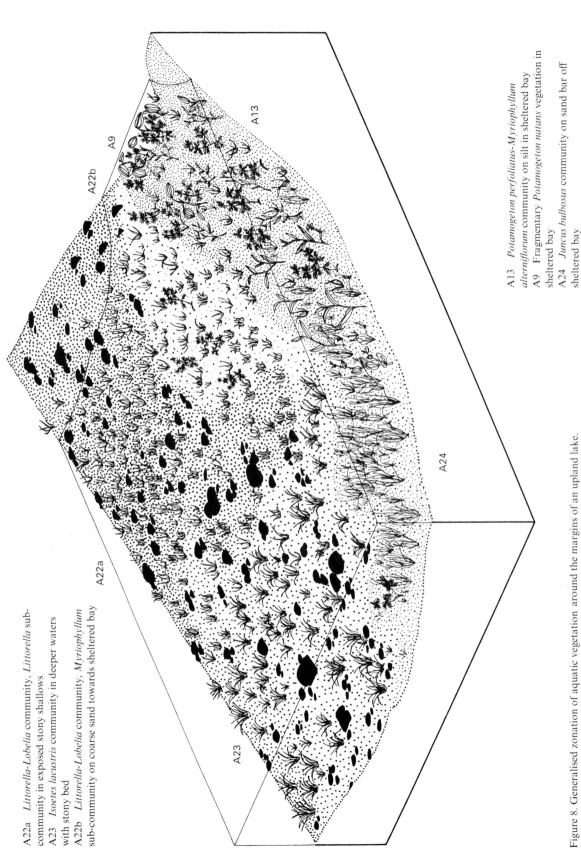

A22a *Littorella-Lobelia* community, *Littorella* sub-community in exposed stony shallows
A23 *Isoetes lacustris* community in deeper waters with stony bed
A22b *Littorella-Lobelia* community, *Myriophyllum* sub-community on coarse sand towards sheltered bay

A13 *Potamogeton perfoliatus-Myriophyllum alterniflorum* community on silt in sheltered bay
A9 Fragmentary *Potamogeton natans* vegetation in sheltered bay
A24 *Juncus bulbosus* community on sand bar off sheltered bay

Figure 8. Generalised zonation of aquatic vegetation around the margins of an upland lake.

SWAMPS AND TALL-HERB FENS

INTRODUCTION TO SWAMPS AND TALL-HERB FENS

The sampling of swamps and tall-herb fens

Swamps are species-poor vegetation types, generally dominated by bulky emergent monocotyledons, characteristic of open-water transitions with permanently or seasonally submerged substrates (Tansley 1939, Spence 1964). However, a much wider variety of communities is included here than has traditionally been encompassed by the British literature in the loosely applied term 'reed-swamp'. As for those fen communities which are also dealt with in this volume, these are vegetation types which would probably be recognised as 'fen' in the traditional British sense (e.g. Pallis 1911, Pearsall 1918, Tansley 1939, Ratcliffe 1977): that is, they are characterised by mixtures of certain of the important swamp emergents and a variety of often tall perennial herbaceous dicotyledons. They are commonly associated with topogenous mires but they are not restricted to open-water transitions and flood-plain systems, nor are they confined to organic substrates (Spence 1964, Wheeler 1975; c.f. Tansley 1939, Ratcliffe 1977). Furthermore, some of the communities included here as fens would probably be considered by certain authors as 'marsh', an expression which is best used informally as a description of any kind of wet ground and/or its vegetation.

These various vegetation types have attracted very uneven attention in Britain. Swamps can be difficult and unpleasant to work in and they reward the persistent with a dismal poverty of species. There is a wealth of information about the ecology of a very few of the important emergents, notably *Phragmites australis* (e.g. Haslam 1970*a*, *b*, *c*, 1971*a*, *b*, 1972*a*, *b*, 1973), *Cladium mariscus* (e.g. Godwin 1929, Godwin & Tansley 1929, Godwin & Bharucha 1932, Conway 1936*a*, 1938, 1942) and *Glyceria maxima* (e.g. Lambert 1946, 1947*a*, *c*, Buttery & Lambert 1965, Westlake 1966). Of the floristic and environmental relationships of many of the remaining swamp dominants, we know next to nothing.

Richer and more diverse fen systems, on the other hand, have long been of interest and there are numerous descriptions of the vegetation of particular localities in,

for example, Broadland (Pallis 1911, Lambert 1946, 1948, 1951, 1965, Wheeler 1978), The Fens (Yapp 1908, Godwin & Tansley 1929, Poore 1956*b*), Breckland (Bellamy & Rose 1961, Haslam 1965), Kent (Rose 1950), Somerset (Willis & Jefferies 1959, Willis 1967), the Cheshire and Shropshire Meres (Clapham in Tansley 1939, Lind 1949, Sinker 1962, *Meres Report* 1980), on the Pennine Carboniferous Limestone (Holdgate 1955*b*, Ingram *et al.* 1959, Sinker 1960, Proctor 1974, Adam *et al.* 1975) and the Durham Magnesian Limestone (Wheeler 1980*d*), and in the Lake District (Pearsall 1918, Tansley 1939, Pigott & Wilson 1978). However, as Wheeler (1980*a*) has pointed out, much of this work reflects the long-standing British interest in hydroseral successions rather than a concern for the systematic description of the vegetation types. Even where detailed diagnoses of plant communities have been provided, it has been difficult to synthesise these into any sort of national framework of floristic variation.

Meanwhile, there has been a gathering pace of interest in characterising whole water bodies and river and ditch lengths in terms of their associated flora, often taking both aquatic macrophytes and emergents into consideration together, sometimes also including bank vegetation. The studies of Haslam (1978, 1982; see also Haslam & Wolseley 1981) and Holmes (1983), for example, have produced a classification of river types with a wealth of information relating the distribution of their plants to distinctive combinations of physical and classical features of beds and waters and the impacts of management. Then, there has been a series of detailed surveys of ditch lengths by the then Chief Scientist Team and England Field Unit of the NCC, on the Somerset Levels and Moors (Wolseley *et al.* 1984), in North Kent (Charman 1981), on the Pevensey Levels (Glading 1980), in Broadland (Reid *et al.* 1989, Doarks 1990, Doarks & Storer 1990, Doarks & Leach 1990, Doarks *et al.* 1990: see also Doarks 1980), North Norfolk (Leach & Reid 1989), the Derwent Ings (Birkinshaw 1991), Essex and Suffolk (Leach & Doarks 1991) and Devon (Leach *et al.* 1991). Continuing long after our own survey work and using a standardised technique that again combined stands of vegetation which we would have sampled separately (Alcock & Palmer 1985), these studies were nevertheless sensitive to the NVC approach (Palmer 1991), and we were ourselves able to make some reference to the results in interpreting our own data on swamps and fens. For the most part, however, we have had to use this work, and the very considerable amount of data originating from NCC surveys of standing water bodies, classified and discussed by site type in Palmer (1989), in a broadly comparative fashion, and not as a source of compatible samples.

Without repeating existing phytosociological surveys or the work of smaller studies that have provided data of the NVC style, we have attempted to obtain a comprehensive coverage of swamp and fen vegetation. As always, samples were situated solely on the basis of the floristic and structural homogeneity of the vegetation, with no particular effort to include stands rich in species or with interesting or rare plants or classic sites with especially striking and varied zonations and mosaics. Indeed, we were concerned to sample widely among the more fragmentary and nondescript swamps and fens that now make up the bulk of our herbaceous wetlands, so as to comprehend the full range of variation among these vegetation types at the present time. In addition, then, to visiting more natural lake sides and river margins, extensive open-water transitions and other fen remnants, sampling was done around artificial ponds and reservoirs, along canals and drainage ditches, in drying and disturbed wetlands, and among industrial and intensive agricultural landscapes.

In larger, more natural swamps and fens, homogeneity was not difficult to detect but, with the kind of rapid transitions seen around open waters, among the fragments of vegetation scattered along busy or managed waterways and in other sorts of disturbed situations, it was often quite hard to select uniform areas for sampling, at least of sufficiently large size. Where possible, quadrats of 4×4 m or 10×10 m were employed but narrow zones had to be sampled using rectangular quadrats of equivalent area and, where stands were smaller than the requisite size, these were sampled in their entirety. Figure 9 illustrates the difference between this approach to such small, fragmentary stands of streams and lakeside swamps and fens and those other survey methods which have used bank lengths, corridors or whole sites as a basis for sampling. Where submerged or floating aquatics occurred in close association with emergents, it was also more sensible to regard these as a structurally independent element of the vegetation and sample and analyse them separately.

As usual, all vascular plants, bryophytes and microlichens were recorded and scored on the Domin scale although, among these vegetation types, the last group of plants was very scarce. Where a mat of algae was a prominent feature, these were recorded as a single taxon. In those few cases where fens were still mown, sampling was timed before the cut and especial care was taken throughout with problematic taxa. Vegetative sedges were sometimes difficult to determine and there may occasionally have been some confusion between *Juncus articulatus* and *J. acutiflorus*. *J. bulbosus/kochii* has been used to accommodate records which did not discriminate between these taxa, and *Rubus fruticosus* and *Nasturtium officinale* were always recorded to the aggregate.

The floristic data were supplemented by observations on the structure of the vegetation: the particular pattern of dominance by the bigger graminoids and bulky sedges, for example, with details of layering or horizontal mosaics, the distribution of associates beneath, between or scrambling among these plants, any suggestions of phenological change with the emergence, growth and death of annual crops of shoots, the impacts of the accumulation and decay of litter, and so on.

Then, to the basic environmental data collected for every sample, were added details of the cover and depth of any open water, an estimate of speed of flow in moving waters, any evidence of seasonal or periodic changes in water depth and flow, signs of landward flushing, or impacts of wave action. Evidence of human impacts, too, was noted, whether on water margins themselves, through dredging, bank maintenance or boat traffic, in draining, disturbance or peat cutting among fens, or through mowing, grazing or burning of the vegetation. Notes were also made on how the stands sampled related to the overall pattern of vegetation on a site, in zonations and mosaics around open water, among extensive fen complexes and where fragments of swamp and fen remained within intensive agricultural landscapes or urban and industrial areas. Any apparent successional processes were noted.

In addition to samples collected by our own research team, we were again fortunate in being allowed access to existing data, though with swamps and fens these were relatively few in number. Some came from studies of particular areas like Skye (Birks 1973) and Malham Tarn (Proctor 1974, Adam *et al.* 1975), with more extensive data for Scotland originating from the work of the Macaulay Institute (Birse & Robertson 1976, Birse 1980, 1984) and for Wales from the Welsh Lowland Peatland Survey (Ratcliffe & Hattey 1982). Studies by Wheeler (1975, 1978, 1980*a*, *b*, *c*) were especially valuable in extending our knowledge and understanding of some of our most striking fens in Broadland and elsewhere.

From these and other smaller papers and unpublished NCC reports, a total of just under 1800 samples was

20 *Scirpus lacustris* ssp. *tabernaemontani* dominant, sometimes with some *S. maritimus*, over a mat of *Agrostis stolonifera* with occasional *Eleocharis palustris*, *E. uniglumis*, *Carex otrubae* or *Oenanthe lachenalii*

S20 *Scirpetum tabernaemontani*
Agrostis stolonifera sub-community

Scirpus lacustris ssp. *tabernaemontani* rare 21

21 *Scirpus maritimus* dominant over a mat of sprawling *Agrostis stolonifera* with scattered *Atriplex prostrata* 22

Scirpus maritimus absent 23

22 *Potentilla anserina* absent

S21 *Scirpetum maritimi*
Agrostis stolonifera sub-community

Potentilla anserina present with occasional *Rumex crispus*, *Festuca arundinacea* and *Cochlearia officinalis*

S21 *Scirpetum maritimi*
Potentilla anserina sub-community

23 *Carex otrubae* prominent in scattered tussocks among a mat of *Agrostis stolonifera*

S18 *Caricetum otrubae*
Atriplex prostrata sub-community

Carex otrubae absent but *Eleocharis palustris* dominant over a mat of *Agrostis stolonifera* and *Potentilla anserina*

S19 *Eleocharitetum palustris*
Agrostis stolonifera sub-community

24 *Phragmites australis* dominant 25

Scirpus maritimus dominant with scattered *Atriplex prostrata* and sometimes with *Puccinellia maritima*

S21 *Scirpetum maritimi*
Atriplex prostrata sub-community

25 Tall and often closed canopy of *Phragmites australis* with scattered *Atriplex prostrata* beneath; *Puccinellia maritima* absent

S4 *Phragmitetum australis*
Atriplex prostrata sub-community
Atriplex prostrata variant

Open and generally short canopy of *Phragmites australis* with scattered *Atriplex prostrata* in a mat of *Puccinellia maritima* and algae

S4 *Phragmitetum australis*
Atriplex prostrata sub-community
Puccinellia maritima variant

26 *Phragmites australis* and *Urtica dioica* always prominent usually with one or more of *Arrhenatherum elatius*, *Oenanthe crocata*, *Calystegia sepium*, *Epilobium hirsutum* in a characteristically patchy tall-herb fen

S26 *Phragmites australis-Urtica dioica* tall-herb fen 27

Not as above 30

27 *Epilobium hirsutum*, *Phragmites australis* and *Urtica dioica* co-dominant or *Glyceria maxima* dominant with these three species abundant

S26 *Phragmites australis-Urtica dioica* tall-herb fen
Epilobium hirsutum sub-community

In more eutrophic waters, it may be difficult to partition samples between this vegetation type and either the *Phragmites australis* sub-community of the *Phragmites-Eupatorium* fen or the *Glyceria maxima* sub-community of the *Peucedano-Phragmitetum*, but the abundance of *Urtica dioica* in the *Phragmites-Urtica* fen is generally a reliable criterion for separation.

Oenanthe crocata and *Calystegia sepium* constant and often abundant

S26 *Phragmites australis-Urtica dioica* fen
Oenanthe crocata sub-community

Arrhenatherum elatius prominent beneath a canopy of *Phragmites australis* and *Urtica dioica* with scattered *Cirsium arvense*

S26 *Phragmites australis-Urtica dioica* fen
Arrhenatherum elatius sub-community

Phragmites australis and *Urtica dioica* co-dominant with *Galium aparine* and *Filipendula ulmaria* constant and sometimes abundant but other tall herbs infrequent

S26 *Phragmites australis-Urtica dioica* fen
Filipendula ulmaria sub-community

28 *Phalaris arundinacea* constant and abundant

S28 *Phalaridetum arundinaceae* 29

Phalaris arundinacea infrequent 30

29 *Urtica dioica* and/or *Epilobium hirsutum* co-dominant with *Phalaris arundinacea*

S28 *Phalaridetum arundinaceae*
Epilobium hirsutum-Urtica dioica sub-community

Grassy understorey with one or more of *Holcus lanatus*, *Elymus repens*, *Poa trivialis* and *Deschampsia cespitosa* present beneath *Phalaris arundinacea* canopy

S28 *Phalaridetum arundinaceae*
Elymus repens-Holcus lanatus sub-community

30   Three or more of *Carex rostrata*, *Menyanthes trifoliata*, *Potentilla palustris* and *Galium palustre* present with one or more of *Calliergon cuspidatum*, *C. cordifolium* and *C. giganteum*

S27 *Carex rostrata-Potentilla palustris* tall-herb fen
Potentillo-Caricetum rostratae   Wheeler 1980*a*
31

Above combination of species absent or, if present, then either *Cladium mariscus* or *Calamagrostis canescens* also present   32

31   *Phragmites australis*, *Lysimachia vulgaris* and *Lythrum salicaria* constant in vegetation usually dominated by *Phragmites australis*, *Carex elata*, *C. nigra*, *Juncus effusus* or *Eriophorum angustifolium*

S27 *Potentillo-Caricetum rostratae*
Lysimachia vulgaris sub-community

In Broadland, it may be difficult to partition samples between this vegetation type and the *Cicuta* sub-community of the *Peucedano-Phragmitetum* but, in general, the high frequency and abundance of *Cladium mariscus* and/or *Calamagrostis canescens* serve to distinguish the latter.

Above combination of species absent; *Equisetum fluviatile* constant and often abundant in vegetation usually dominated by *Carex rostrata*, *Potentilla palustris*, *Menyanthes trifoliata*, *Equisetum fluviatile*, *Juncus effusus*, *J. acutiflorus*, *Carex vesicaria* or *C. aquatilis*

S27 *Potentillo-Caricetum rostratae*
Carex rostrata-Equisetum fluviatile sub-community

Around lake margins to the north and west, this vegetation type may grade imperceptibly to swamps of the *Caricetum rostratae*, the *Equisetum fluviatile*, the *Caricetum vesicariae* or to virtually pure stands of *Menyanthes trifoliata* or *Potentilla palustris*.

---

32   Six or more of the following present in often species-rich vegetation with complex structure and very variable pattern of dominance: *Phragmites australis*, *Calamagrostis canescens*, *Cladium mariscus*, *Carex elata*, *Juncus subnodulosus*, *Galium palustre*, *Lythrum salicaria*, *Lysimachia vulgaris*, *Peucedanum palustre*, *Eupatorium cannabinum*, *Filipendula ulmaria*, *Mentha aquatica*, *Calliergon cuspidatum*, *Campylium stellatum*

S24 *Phragmites australis-Peucedanum palustre* tall-herb fen
Peucedano-Phragmitetum australis Wheeler 1978 emend.   33

Simpler and less species-rich vegetation generally lacking combinations of the above species and certainly with *Calamagrostis canescens*, *Lysimachia vulgaris*, *Peucedanum palustre* and *Campylium stellatum* rare

S25 *Phragmites australis-Eupatorium cannabinum* tall-herb fen   39

33   *Carex paniculata* present and sometimes dominant in the virtual absence of *Cladium mariscus*   34

Carex paniculata very infrequent but *Cladium mariscus* frequent and often dominant   35

34   *Glyceria maxima* constant and often dominant with *Epilobium hirsutum*, *Carex riparia*, *C. acutiformis* sometimes locally abundant

S24 *Peucedano-Phragmitetum australis*
Glyceria maxima sub-community

Along the Yare valley in Norfolk, it may be difficult to partition samples between this vegetation type and either the *Epilobium hirsutum* sub-community of the *Phragmites-Urtica* fen or the *Caricetum ripariae*, but the frequency of *Thalictrum flavum* and *Lathyrus palustris* in this kind of *Peucedano-Phragmitetum* will usually serve as a diagnostic criterion.

Glyceria maxima infrequent and never abundant in vegetation dominated by tussocks of *Carex paniculata*

S24 *Peucedano-Phragmitetum australis*
Carex paniculata sub-community

35   *Symphytum officinale* frequent in small amounts often with *Phalaris arundinacea*, *Thalictrum flavum* and *Lathyrus palustris* in rank vegetation generally dominated by *Phragmites australis*, *Cladium mariscus*, *Calamagrostis canescens* or *C. epigejos*

S24 *Peucedano-Phragmitetum australis*
Symphytum officinale sub-community

Symphytum officinale absent   36

36   *Schoenus nigricans* constant, sometimes with *Oenanthe lachenalii*, in vegetation generally dominated by *Cladium mariscus* or *Phragmites australis*

S24 *Peucedano-Phragmitetum australis*
Schoenus nigricans sub-community

Schoenus nigricans infrequent   37

37   Scattered bushes of *Myrica gale* present often with some creeping *Salix repens* in vegetation dominated by *Phragmites australis, Cladium mariscus* or *Calamagrostis canescens*

S24 *Peucedano-Phragmitetum australis*
Myrica gale sub-community

Myrica gale and *Salix repens* infrequent   38

38   Three of more of *Cicuta virosa, Ranunculus lingua, Carex pseudocyperus, Berula erecta, Sium latifolium* present in vegetation dominated by *Phragmites australis, Carex lasiocarpa, C. paniculata* or *C. riparia*

S24 *Peucedano-Phragmitetum australis*
Cicuta virosa sub-community

Combination of the above species infrequent in vegetation dominated by *Phragmites australis* or *Calamagrostis canescens*

S24 *Peucedano-Phragmitetum australis*
Typical sub-community

39   *Cladium mariscus* constant and often co-dominant with *Phragmites australis* or, less frequently, *Carex elata*

S25 *Phragmites australis-Eupatorium cannabinum* tall-herb fen
Cladium mariscus sub-community

Some richer stands of this vegetation type may be difficult to separate from poorer stands of the Typical sub-community of the *Peucedano-Phragmitetum* but the absence here of *Calamagrostis canescens, Lysimachia vulgaris* and *Peucedanum palustre* should serve as a diagnostic criterion.

Cladium mariscus rare and never dominant   40

40   *Carex paniculata* constant and usually dominant with little or no *Juncus subnodulosus* beneath

S25 *Phragmites australis-Eupatorium cannabinum* tall-herb fen
Carex paniculata sub-community

Some richer stands of this vegetation type may be difficult to separate from poorer stands of the *Carex paniculata* sub-community of the *Peucedano-Phragmitetum* but the absence here of *Lysimachia vulgaris, Peucedanum palustre, Scutellaria galericulata* and *Impatiens capensis* should serve as a diagnostic criterion.

Phragmites australis usually dominant with *Carex paniculata* rare and never abundant

S25 *Phragmites australis-Eupatorium cannabinum* tall-herb fen
Phragmites australis sub-community

Some richer stands of this vegetation type may be difficult to separate from poorer stands of the Typical sub-community of the *Peucedano-Phragmitetum* but the absence here of *Calamagrostis canescens, Lysimachia vulgaris* and *Peucedanum palustre* should serve as a diagnostic criterion. Stands of this vegetation type rich in *Epilobium hirsutum* may also be difficult to separate from the *Epilobium hirsutum* sub-community of the *Phragmites-Urtica* tall-herb fen but *Urtica dioica* is rare here and never abundant.

COMMUNITY DESCRIPTIONS

S1
Carex elata sedge-swamp
Caricetum elatae   Koch 1926

Synonymy

Carex elata consocies   Pearsall 1918; Open carr   Pearsall 1918 *p.p.*; *Carex elata* associations Holdgate 1955*b p.p.*; *Carex elata* headwater fen community Haslam 1965; Association of *Carex elata*   Pigott & Wilson 1978.

Constant species

Carex elata.

Rare species

Calamagrostis stricta.

Physiognomy

The *Caricetum elatae* comprises vegetation dominated by *Carex elata*, usually as prominent tussocks. These are up to 40 cm in diameter and height, occasionally taller in deeper water, are often closely up set with to 60 tussocks per 10 × 10 m (Wheeler 1975) and have a canopy of spreading leaves about 1 m long. The community is generally species-poor. Most frequently, there are taller herbaceous dicotyledons scattered around the tussocks and often rooted in the peaty stocks at or around water-level; among these, *Cirsium palustre*, *Eupatorium cannabinum*, *Lycopus europaeus* and *Ranunculus lingua* are the most frequent. *Galium palustre* and *Solanum dulcamara* sometimes form thick tangles of sprawling shoots. In areas of shallow water between the tussocks, scattered plants of *Menyanthes trifoliata*, *Mentha aquatica*, *Potentilla palustris* and *Berula erecta* may occur and, in deeper pools, there may be elements of floating-leaved or submerged aquatic vegetation. In deeper water, too, *Cladium mariscus* may be locally prominent, its clumps forming mosaics with the *C. elata* (Haslam 1965, Wheeler 1975). Occasionally, scattered shoots of *Phragmites australis* form a sparse cover.

The *C. elata* tussocks themselves, especially those which are less vigorous, form a distinctive support for epiphytic *Hydrocotyle vulgaris*, *Epilobium palustre*, *Cardamine pratensis* and (e.g. Holdgate 1955*b*) *Filipendula ulmaria*. Here, too, there may also be sparse wefts of *Calliergon cuspidatum* but, in general, bryophytes are rare.

Saplings of *Salix cinerea* or, in the north, *S. atrocinerea*, are a frequent feature of the community. These seem to gain a hold mostly between the tussocks, on more open areas of bare organic matter (Haslam 1965, Pigott & Wilson 1978). *Alnus glutinosa* is a less frequent invader but it may be able to colonise the tussock tops.

Habitat

The community usually occurs as emergent vegetation in up to 40 cm of water in shallow pools and old peat cuttings, more fragmentarily in derelict ditches (Wheeler 1975) and sometimes as part of open-water transitions around larger water bodies (Pearsall 1918, Holdgate 1955*b*, Pigott & Wilson 1978). Some of the best stands are found in the hollows of what seem to be collapsed pingos or ground-ice depressions on the west Norfolk commons (Sparks *et al.* 1972). In these localities, Haslam (1965) has suggested that *C. elata* dominance is particularly associated with a combination of generally waterlogged conditions and a fluctuating water-table. The north-western localities of the community are also characterised by these features.

The tussocks are commonly rooted in the somewhat unconsolidated organic material over the base of the swamp basins and, in deeper water, they may form a highly unstable semi-floating mat, held together by a tangle of interwoven lateral roots just below the water surface (Wheeler 1975). More rarely, the tussocks are rooted directly in the underlying mineral substrate. In the west Norfolk hollows, organic material lies over thin smears of glacial gravels on Chalk and the water pH ranges from 5.5 to 7.2.

Where it is accessible to stock (a rare occurrence), *C. elata* seems to be grazed (Holdgate 1955*b*).

Zonation and succession

In west Norfolk, the *Caricetum elatae* may occasionally fill more shallow hollows but it usually occurs as a zone around open water or gives way in deeper water to the

Cladietum marisci (Haslam 1965, Wheeler 1975). Mosaics of the two dominants commonly occur at the junctions of the communities. Away from the water, the zonation varies with the basin slope: on more gradual surrounds, there is often a fairly sharp transition to some kind of rush-pasture and then to calcicolous grassland; around more steeply sided hollows, calcicolous grassland may abut directly on to the *Caricetum elatae* (Wheeler 1975).

In its Cumbrian localities, the community occurs in close association with the *Potentillo-Caricetum rostratae* to which it grades in deeper water and elements of which may occur between the *C. elata* tussocks forming mosaics (Pearsall 1918, Holdgate 1955b, Pigott & Wilson 1978).

The advance of the *Caricetum elatae* and its gradual colonisation by *Salix atrocinerea* and *Alnus glutinosa* form part of the successional changes catalogued at Esthwaite North Fen over some 50 years (Pearsall 1918, Tansley 1939, Pigott & Wilson 1978). Here, an area of what seems to have been *Phragmitetum australis* reed-swamp and *Potentillo-Caricetum rostratae* in 1914–16 was occupied by the *Caricetum elatae* by 1929 and by 1969 had an open woodland cover. *C. elata* appears to be tolerant of a certain amount of shade (Pigott & Wilson 1978), although beneath a *Salix-Alnus* canopy, its dominance weakens and the major tussock sedge in the open woodland at this site is *C. paniculata*.

Distribution

The *Caricetum elatae* is an uncommon community, being restricted to a few localities in west Norfolk, Anglesey and Cumbria. The Scottish locality on Loch Ness described by West (1905) has not been visited but it had a predominance of *Carex rostrata* when re-surveyed by Spence (1964).

Affinities

C. elata is a sedge with a predominantly eastern distribution in Britain (Jermy *et al.* 1982) and in this country it also occurs as a component, and occasionally as a dominant, in the *Peucedano-Phragmitetum australis* and *Phragmites australis-Eupatorium cannabinum* fens and, much less commonly, in the *Potentillo-Caricetum rostratae* and the *Carex rostrata-Calliergon* fens. In this scheme, the *Caricetum elatae* is retained to characterise those very distinctive situations where *C. elata* occurs as the sole dominant of species-poor vegetation in the general absence of large amounts of all other swamp and fen dominants. These are much less common in Britain than in Continental Europe.

The scope of the community is somewhat narrower than that originally proposed by Koch (1926) and developed by certain European authors (e.g. Vollmar 1947) and the vegetation described here seems to be akin to that placed in a more species-poor sub-Association,

comaretosum palustris (= *potentilletosum palustris*), by van Donselaar (1961) and Westhoff & den Held (1969). *C. elata* swamps with *Phragmites australis* or *Cladium mariscus* have sometimes been separated off into distinct vegetation types, e.g. the *Caricetum elatae phragmitosum* (Koch 1926, Vollmar 1947) and the *Caricetum elatae cladietosum* (Libbert 1932).

Floristic table S1

Carex elata	V (7–9)
Cirsium palustre	III (1–3)
Galium palustre	III (1–3)
Hydrocotyle vulgaris	III (1–3)
Eupatorium cannabinum	III (1–3)
Juncus subnodulosus	III (1–3)
Potentilla palustris	III (1–5)
Salix cinerea saplings	III (1–3)
Solanum dulcamara	III (1–3)
Calliergon cuspidatum	III (1–5)
Epipactis palustris	III (1–3)
Menyanthes trifoliata	III (1-3)
Lycopus europaeus	III (1–3)
Ranunculus lingua	III (1–3)
Cladium mariscus	II (1–5)
Phragmites australis	II (1–3)
Equisetum palustre	II (1–3)
Mentha aquatica	II (1)
Caltha palustris	II (1–3)
Berula erecta	II (1–5)
Lythrum salicaria	I (1)
Epilobium palustre	I (4)
Carex nigra	I (4)
Juncus acutiflorus	I (2)
Agrostis canina canina	I (5)
Anthoxanthum odoratum	I (4)
Viola palustris	I (4)
Brachythecium rutabulum	I (4)
Holcus lanatus	I (1)
Cardamine pratensis	I (4)
Rhytidiadelphus squarrosus	I (1)
Rumex acetosa	I (1)
Eurhynchium praelongum	I (1)
Number of samples	7
Number of species/sample	12 (6–17)

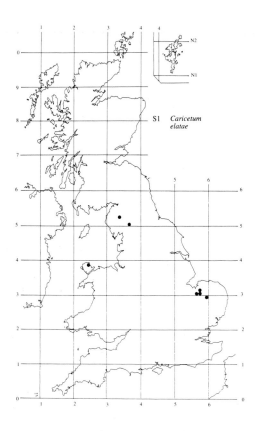

S1   *Caricetum elatae*

S2
Cladium mariscus swamp and sedge-beds
Cladietum marisci   Zöbrist 1933 *emend*. Pfeiffer 1961

Synonymy
Pure Sedge Godwin & Tansley 1929; *Cladietum*
Tansley 1939; *Cladium mariscus* reed-swamp Conway
1942; *Cladium mariscus* primary fen Lambert 1951;
Cladium mariscus stands Holdgate 1955*b*; *Cladium
mariscus* society Poore & Walker 1959; Tall *Cladium
mariscus* Haslam 1965; *Cladium mariscus* swamp
Ratcliffe & Hattey 1982.

Constant species
Cladium mariscus.

Rare species
Utricularia intermedia.

Physiognomy
The *Cladietum marisci* comprises vegetation overwhelmingly dominated by *Cladium mariscus*. The gregarious stout shoots up to 2 metres tall form a sometimes
open patchwork of clumps or a dense cover, often made
virtually impenetrable by the tough leaves which are up
to 3 m long and characteristically bent over, 1 m or more
from the ground. This growth habit, the evergreen
nature of the aerial parts and the thick accumulation of
slow-rotting litter, sometimes more than 50 cm deep
(Godwin & Tansley 1929, Conway 1942), can exclude all
possible competitors. No other species is frequent
throughout and pure stands are common.

Sub-communities

***Cladium mariscus* sub-community:** *Scirpo-Phragmitetum
cladietosum* Messikommer 1928; *Caricetum elatae cladietosum* Libbert 1932; *Cladium marisci typicum* Krausch 1964 *p.p.*; *Cladietum marisci phragmitetosum* Westhoff & Segal 1969; *Cladietum marisci
typicum, Cladium-Carex elata* community, *Cladium-
Thelypteris* community, *Cladium* sociation Wheeler
1980*a*. Here are included the most dense and/or species-
poor stands, ranging from pure *Cladium*, through more

open vegetation with occasional shoots of *Phragmites
australis* or *Juncus subnodulosus* and sprawls of *Solanum
dulcamara* to mosaics of *C. mariscus* with sometimes
abundant *P. australis* or scattered tussocks of *Carex
elata.*

***Menyanthes trifoliata* sub-community:** *Cladium maris-
cus-Myrica gale* sociation Spence 1964; *Cladium mariscus-Utricularia intermedia* nodum Ivimey-Cook &
Proctor 1966; *Cladietum marisci scorpidietosum* Segal
& Westhoff 1969; *Cladietum marisci utricularietosum*
and *caricetosum lasiocarpae* Wheeler 1980*a*; *Cladium
mariscus-Rubus fruticosus-Myrica gale* community
Ratcliffe & Hattey 1982. In this sub-community, the
cover of *Cladium* is more open and the plants often
shorter, up to about 1 m tall. *P. australis, J. subnodulus*
and *C. elata* are more frequent here and there are
sometimes scattered clumps of *Myrica gale* but the really
distinctive feature is the vegetation of the pools of
standing water between the *Cladium* shoots. *Menyanthes
trifoliata* is the most frequent and abundant species but
Potentilla palustris* and, especially in shallower water,
Carex lasiocarpa, may be locally prominent. Submerged, there are sometimes festoons of *Scorpidium
scorpioides* and *Utricularia vulgaris* with, less frequently,
U. minor, U. intermedia and *Campylium stellatum*.
Wheeler (1980*a*) also recorded *Chara vulgaris, C. glubu-
laris* var. *aspera* and *C. hispida* var. *hispida* as distinctive
aquatics of this vegetation.

Habitat
The *Cladietum* is most characteristic of shallow, standing water in lowland topogenous mires fed by calcareous, base-rich ground water. It occurs, often as small
stands, in basin mires and around open-water transitions in larger hollows and flood-plain mires. It may also
be found in flooded peat cuttings and around
'pulk-holes'.
    The community reaches its optimal development as a
swamp. *Cladium* is an evergreen geophyte which, in

many respects, can be regarded as an aquatic species (Conway 1936a). It grows best when the water-table remains between about 15 cm below ground and 40 cm above (Conway 1942). Although it can tolerate short periods of drier conditions in summer (e.g. Godwin & Bharucha 1932), vigour is maintained only if the dense mat of fleshy roots is submerged (Conway 1937b, 1938). On the other hand, a persistent very high water-table can prevent aeration of the rhizome which occurs by gaseous diffusion down the bases of the dead leaves and through those which, though still green, have ceased growth (Conway 1937a). These requirements probably play an important part in limiting the species to situations which are 'too dry for reed and too wet for bushes' (Godwin & Tansley 1929). Swamp dominated by *Phragmites australis*, in which the rhizomes are aerated through the very tall, persistent dead stems (Buttery 1959, Haslam 1972), commonly replaces the *Cladietum* in deeper water in uninterrupted open-water transitions (e.g. Godwin & Bharucha 1932).

Between these general limits, *Cladium* seems relatively tolerant of the actual level of the water-table except that, in winter, standing water may provide some insulation against frosting of the growing point (which is injured by temperatures below $-2\,°C$) and the differentiated leaves (which can withstand temperatures down to $-10\,°C$) (von Post 1925, Conway 1938, 1942). It is noteworthy that, to the north of Britain, *Cladium* occurs only in the swamp *Cladietum* and not as a component of drier fens (Wheeler 1975).

It has, however, been suggested (Haslam 1965) that part of the balance between *Cladium* and *Carex elata*, both of which can occur in broadly similar situations, is related to tolerance of short-term variation in water-level. In the west Norfolk basin mires, *C. elata* is limited to sites with a variable water-table, whereas *Cladium* is dominant where the water-table remains more stable. Wheeler (1980a) noted that Broadland stands of the *Cladietum* with *C. elata* had a more variable water-table than those from which this sedge was lacking.

The *Cladietum* occurs mostly over fen peat, although it is sometimes found over marly substrates (Ratcliffe & Hattey 1982) and may even colonise bare silts, as around Broadland 'pulk-holes' (Lambert 1951), although severe clogging with mineral material may inhibit rhizome aeration (Haslam 1965). The community is commonly associated with infiltration of calcareous and base-rich water, sometimes by general seepage from surrounding rocks (as in the mires in limestone basins), in other cases by more localised infiltration from springs. Calcium contents of 40–150 mg l⁻¹ have been recorded (Poore & Walker 1959, Phillips 1977, Wheeler 1983) with water pH values of 6.0–8.5 (Godwin & Bharncha 1932, Conway 1942, Poore & Walker 1959, Wheeler 1975, 1983). There is, however, some evidence

to suggest that *C. mariscus* is not always restricted to such conditions throughout its European range (e.g. Praeger 1934, Skårman 1935).

Cladium may gain some competitive advantage over, for example, *Phragmites*, by a tolerance of the oligotrophic environment that is sometimes associated with such calcareous conditions. In her study of the successions along the Bure valley in Norfolk, Lambert (1951) noted that the importance of the *Cladietum* in the vegetation sequences tended to increase the further the Broads basins were from the river and she suggested that this might be related to a decreased throughput of nutrients. A recent study of Upton Broad (Phillips 1977), where the community is particularly well developed at a site far from the river and also land-locked, has shown not only high levels of calcium but also low amounts of nitrate-nitrogen and phosphate. In the west Norfolk valley mires, *Cladium* is generally absent from those sites where frequent inundation with nutrient-rich mineral sediments allows *Phragmites* to thrive (Haslam 1965).

Stands of the *Cladietum* can provide a good crop of virtually pure 'sedge' for thatch but frequent cutting can modify the floristics of the vegetation. Traditionally, at Wicken and more widely in Broadland, *Cladium* sedge beds were mown in early summer. (T. Rowell, personal communication: cf. Godwin & Tansley 1929), every 3 to 5 years depending on the condition of the ground, and such treatment appears to be optimal for the renewable management of the crop.

With its thick accumulation of dead and decaying leaves beneath, the community is very susceptible to burning when dry. Indeed, dried *Cladium* leaves at one time had a ready use as kindling.

Zonation and succession
The centres of small basin mires and peat cuttings may be completely filled by dense stands of the community. In other cases, the *Cladietum* survives patchily in pools and stagnant dykes within largely terrestrialised mire complexes, giving way to a variety of fen or poor-fen communities. In grossly disturbed and abandoned sites, the community may occur in juxtaposition with the *Phragmites australis-Urtica dioica* fen.

In larger hollows, the *Cladietum* may give way directly to open standing water or, around 'pulk-holes', it may abut on to bare silt (Lambert 1951). In some sites, there is a swamp mosaic of the community with the *Caricetum elatae*, apparently in relation to the height and short-term variability of the water-table (Haslam 1965). Around more extensive open-water transitions, the *Cladietum* may occur patchily in the shallower waters in association with *Phragmites* swamps and, especially to the north and west, the *Caricetum rostratae* (Holdgate 1955b). Exceptionally the community may be found as part of more intact zonations as at Upton Broad in Norfolk. Here, Lambert & Jennings (1951) described sequences running from the *Typhetum angustifoliae* and/or *Phragmitetum australis* through the *Cladietum* to carr woodland on the thicker, drier peats away from the open broad.

From the more fragmentary pattern of vegetation in the flood-plain mire remnants at Wicken Fen, Godwin & Bharucha (1932) had shown how the distribution of *Phragmites* swamp, the *Cladietum* and carr in such transitions could be related to differences in the height of the water-table, and particularly to the extent of water excess in winter. This they saw as representing a primary succession involving the colonisation of open water by *Phragmites*, still remaining as a swamp along lodes and around pools, the invasion of the *Phragmites* swamp by *Cladium*, best seen in the old peat cuttings, and the gradual colonisation of the 'Pure Sedge' by bushes. Stratigraphical analysis of the Broadland deposits under the zonations described by Lambert & Jennings (1951) revealed a similar picture, except that there, a second phase of *Phragmites* dominance seemed to be interposed between the *Cladietum* and the carr (Lambert 1951).

Despite this, and the rarity with which Lambert (1951) observed direct bush invasion of dense uncut *Cladietum* around the Bure broads, the advance of woody species into the community at Wicken has been plotted in some detail (Godwin & Tansley 1929, Godwin & Bharucha 1932). Here, the most frequent invaders were *Frangula alnus*, then *Rhamnus catharticus*, *Salix cinerea* and *Viburnum opulus*, with very small amounts of *Crataegus monogyna* and *Betula pendula*. These gained a hold in the *Cladietum* towards its uppermost boundary, the seedlings surviving down to a critical limit of bush growth where the peat surface experienced shallow winter flooding of several weeks' duration (Godwin & Bharucha 1932, Godwin 1943b). Initial patchy bush colonisation over a period of 20 years or so was followed by gradual infill of the canopy and eventual extinction of the dense *Cladium* cover (Godwin & Tansley 1929, Godwin 1936). The long survival of the sedge was perhaps due to its evergreen nature (Conway 1942).

Studies at Wicken also revealed how repeated summer cutting of the *Cladietum* 'Pure Sedge' might deflect the primary succession to produce 'Mixed Sedge' and 'Litter' (Godwin 1929, 1941; Godwin & Tansley 1929), vegetation represented in this scheme by various kinds of *Cirsio-Molinietum*, described among the mires of Volume 2.

Distribution
The community has a local distribution within the British range of *Cladium* which, as elsewhere in Europe, is benefited by warm summers and the absence of

intensive frost (von Post 1925, Conway 1938, 1942). It is best developed in calcareous basin mires which are rather uncommon in this country (Wheeler 1983) but well seen on the Carboniferous Limestone of Anglesey (Wheeler 1975, Ratcliffe & Hattey 1982) and on the west Norfolk commons where drift smears overlie Chalk (Haslam 1965, Wheeler 1975). More isolated occurrences of this kind are on the Carboniferous Limestone of Cumbria, as at Sunbiggin Tarn (Holdgate 1955b), where *Cladium* reaches its altitudinal limit of 260 m in Britain, and at Hell Kettles, small subsidence hollows in Magnesian Limestone in south Durham (Wheeler 1980d). Other scattered localities sometimes mark localised seepage of calcareous water, as on Wybunbury Moss in Cheshire (Poore & Walker 1959) and on Anabaglish Moss in Wigtownshire (Spence 1964).

The *Cladietum* also occurs around open-water transitions in the apparently more oligotrophic flood-plain mires, as along the Bure (Lambert 1951) and in peat-cuttings, such as the celebrated sites at Wicken Fen (Yapp 1908, Godwin & Tansley 1929).

Affinities
Towards southern Britain, *Cladium* is a component, and sometimes an important dominant, in a variety of fens, most notably the *Peucedano-Phragmitetum australis* and the *Phragmites australis-Eupatorium cannabinum* fens and, less frequently, the *Phragmites australis-Urtica dioica* fen and the *Potentillo-Caricetum rostratae*. As with the *Caricetum elatae*, the *Cladietum marisci* is here retained in a rather strict sense, to contain those species-poor stands in which *Cladium* is dominant in the general absence of large amounts of other important swamp species.

This treatment approximates most nearly to the revision of the *Cladietum marisci* of Zöbrist (1935) proposed by Pfeiffer (1961) and followed by Wheeler (1975, 1980a). In this view, the community is essentially a species-poor swamp, although it contains some vegetation in which other possible swamp dominants, such as *Carex elata*, are locally prominent and thus takes in such communities as the *Caricetum elatae cladietosum* of Libbert (1932). In addition, the *Cladietum* as understood here includes some stands with much *Phragmites* which Wheeler (1975, 1980a), following Pfeiffer (1961), retained in a *Cladium*-rich type of *Scirpo-Phragmitetum* (similar to the *Scirpo-Phragmitetum cladietosum*) of Messikommer (1928)).

Such an approach contrasts with that of Krausch (1964) and Oberdorfer (1965) who, alongside a species-poor swamp core, included within the *Cladietum* some rich-fen vegetation that is here grouped in the *Peucedano-Phragmitetum* and *Phragmites-Eupatorium* communities.

Floristic table S2

	a	b	2
Cladium mariscus	V (5–10)	V (7)	V (5–10)
Calliergon cuspidatum	II (1–5)		I (1–5)
Solanum dulcamara	II (1–5)		I (1–5)
Salix cinerea sapling	I (2–3)		I (2–3)
Fissidens adianthoides	I (1)		I (1)
Equisetum palustre	I (1)		I (1)
Phragmites australis	II (1–6)	V (1–8)	III (1–8)
Menyanthes trifoliata	I (1–3)	IV (1–8)	II (1–8)
Potentilla palustris		II (1–5)	I (1–5)
Carex lasiocarpa	I (5)	II (1–5)	I (1–5)
Scorpidium scorpioides	I (1–3)	II (1–3)	I (1–3)
Utricularia vulgaris		II (1–3)	I (1–3)
Mentha aquatica		II (1–5)	I (1–5)
Utricularia minor		I (1–3)	I (1–3)
Utricularia intermedia		I (1–3)	I (1–3)
Riccardia multifida		I (1)	I (1)
Campylium stellatum		I (1–5)	I (1–5)
Hippuris vulgaris		I (1)	I (1)
Sphagnum subnitens		I (1)	I (1)
Carex panicea		I (1)	I (1)
Caltha palustris		I (1)	I (1)
Carex rostrata		I (1)	I (1)
Juncus subnodulosus	II (1–5)	III (1–8)	III (1–8)
Galium palustre	I (2)	II (1–3)	I (1–3)
Carex elata	I (3)	II (3)	I (3)
Myrica gale	I (8)	II (5)	I (5–8)
Lythrum salicaria	I (1)	I (1)	I (1)
Number of samples	30	24	54
Number of species/sample	5 (1–12)	9 (7–10)	7 (1–12)

a   *Cladium mariscus* sub-community
b   *Menyanthes trifoliata* sub-community
2   *Cladietum marisci* (total)

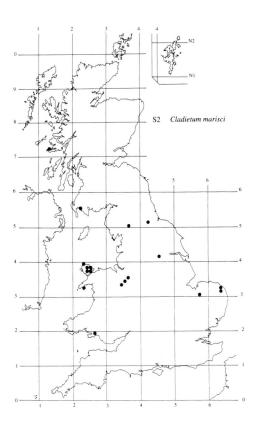

S2   *Cladietum marisci*

S3
Carex paniculata sedge-swamp
Caricetum paniculatae Wangerin 1916

Synonymy

Caricetum paniculatae Tansley 1939; Primary tussock fen Lambert 1951 *p.p.*; *Carex paniculata* swamp Poore & Walker 1959, Sinker 1962; *Carex paniculata-Angelica sylvestris* sociation Spence 1964 *p.p.*; *Caricetum paniculatae typicum* Wheeler 1980a; *Carex paniculata-Rubus fruticosus* community Ratcliffe and Hattey 1982.

Constant species

Carex paniculata.

Physiognomy

The *Caricetum paniculatae* is dominated by tussocks of *Carex paniculata*, the stocks of which may attain a massive size, often reaching more than 1 m in height and diameter and being crowned by spreading stems and leaves 1 m or so long. Between the tussocks, which, at one site (Sweat Mere, Shropshire: Clapham in Tansley 1939), had centres 0.75–3 m apart, there is standing water or exposed peat and silt. Here, the vegetation is characteristically sparse and species-poor. There may be a few shoots of emergent *Phragmites australis*, *Sparganium erectum*, *Typha latifolia*, *Equisetum fluviatile* or *Epilobium hirsutum* and, beneath scattered plants of *Caltha palustris*, *Viola palustris* and *Myosotis scorpioides* and sometimes *Potentilla palustris* and *Menyanthes trifoliata*; in other cases, just a few wefts of *Eurhynchium praelongum* and *Brachythecium rutabulum* and occasional plants of *Lemna minor* occur on largely bare expanses of substrate.

The tussocks themselves usually support some epiphytes, although the flora here is never as rich as that on *C. paniculata* tussocks in the *Peucedano-Phragmitetum* and the *Phragmites australis-Eupatorium cannabinum* fen. In particular, seedlings of *Salix cinerea* or, in the north, *S. atrocinerea*, and *Alnus glutinosa*, are rare. However, some of the following are generally present: *Angelica sylvestris*, *Filipendula ulmaria*, *Galium palustre*, *Rubus fruticosus* agg., *Solanum dulcamara* and, particu-

larly distinctive, *Athyrium filix-femina* and *Dryopteris dilatata*.

Habitat

The community is most characteristic of the shallows of lowland open-water transitions such as occur around lakes, pools and abandoned ox-bows. Less frequently, it may be found in basin, valley and flood-plain mires and in peat cuttings. *C. paniculata* is often associated with situations where there is some, at least seasonal, movement in and eutrophication of base-rich waters (e.g. Lambert 1951). Here, however, it is dominant in waters that, though often calcareous and base-rich (calcium content 71–74 mg l^{-1}, pH 7.1–8.1: Poore & Walker 1959, Sinker 1962), have little through-put and are perhaps more mesotrophic.

Stands usually occur on a base of semi-fluid to firm but floating *P. australis* or *Typha angustifolia* peat which may become depressed as the tussocks enlarge. Although standing water may thus occur between the tussocks, the water-table in general remains fairly stable around the substrate surface.

On firmer ground, accessible stands may be grazed and battered by stock and the surface between the tussocks badly poached.

Zonation and succession

The *Caricetum paniculatae* may abut directly on to open water or give way to the *Phragmitetum australis* or the *Typhetum angustifoliae*. Although many stands are small and zonations fragmentary, the community is sometimes part of an intact sequence of vegetation from such deeper water swamps to carr (Clapham in Tansley 1939, Lambert 1951, Sinker 1962). In such cases, the *Caricetum paniculatae* may represent the most species-poor swamp phase of *C. paniculata* dominance and give way to woodland through types of richer fen in which the sedge is still very prominent but where there is a more advanced colonisation of older tussock tops by *Salix cinerea* or *S. atrocinerea* and *Alnus glutinosa*, as in the communities of the *Peucedano-Phragmitetum* and the *Phragmites-Eupatorium* and *Phragmites-Urtica* fens. There is strong circumstantial, or sometimes firm stratigraphical, evidence for regarding such zonations as representing a seral development from open water to woodland in which the *Caricetum paniculatae* plays a major part (Clapham in Tansley 1939, Lambert 1951, Lambert & Jennings 1951, Sinker 1962, Wheeler 1983).

Around pools in pasture, such sequences as those described may terminate above in an abrupt transition from the *Caricetum paniculatae* to some kind of grazed Calthion grassland within which tree colonisation is repeatedly set back.

Where the community occurs more patchily in response to local infiltration of calcareous, base-rich waters into basin mires, it may be surrounded by a variety of poor fens or base-poor mire communities (e.g. Poore & Walker 1959).

Distribution

The community is widespread but local. Particularly good examples occur around the open-water transitions of the Shropshire meres (e.g. Clapham in Tansley 1939, Sinker 1962) and in Pembrokeshire valley and floodplain mires (Ratcliffe & Hattey 1982).

Affinities

C. paniculata is a sedge of fairly widespread distribution throughout the English and Welsh lowlands and, less extensively, in Scotland (Jermy *et al.* 1982). Its most renowned occurrences are in richer fens where, though dominant, it occurs with a variety of other important swamp and fen species and a wide range of associates, as in the *Peucedano-Phragmitetum* and the *Phragmites-Eupatorium* communities and, less commonly, the *Phragmites-Urtica* fen and the *Potentillo-Caricetum rostratae*. Here, the *Caricetum paniculatae* is retained to include species-poor swamp stands in which the sedge is overwhelmingly dominant in the virtual absence of other possible swamp and fen dominants.

Some authorities have placed such vegetation within the *Caricetum acutiformo-paniculatae* Vl. & van Zinderen Bakker 1942 where *Carex acutiformis* and *C. riparia* are the usual dominants. In this scheme, a distinct syntaxon in the sense of the *Caricetum paniculatae* Wangerin 1916 is preferred. This was a course adopted by Wheeler (1975, 1980*a*), although the community as diagnosed here includes only his *typicum* sub-community. Wheeler's *peucedanetosum* (Wheeler 1978, 1980*a*) is here placed alongside other very rich fens in an expanded *Peucedano-Phragmitetum*.

Floristic table S:

Carex paniculata	
Angelica sylvestris	
Athyrium filix-fem	
Filipendula ulmari	
Galium palustre	
Rubus fruticosus	
Solanum dulcam	
Epilobium hirsut	
Dryopteris dilat	
Rumex acetosa	
Molinia caerule	
Typha latifolia	
Viola palustris	
Equisetum fluvi	
Phragmites austra	
Caltha palustris	I (1–5)
Oenanthe crocata	I (3–5)
Potentilla palustris	I (5)
Urtica dioica	I (1–3)
Eupatorium cannabinum	I (1)
Epilobium palustre	I (2)
Scutellaria galericulata	I (3)
Cirsium palustre	I (1–3)
Eurhynchium praelongum	I (1–5)
Brachythecium rutabulum	I (1–5)
Sparganium erectum	I (1–3)
Galium aparine	I (1–3)
Myosotis scorpioides	I (2)
Lycopus europaeus	I (3)
Lotus uliginosus	I (2)
Rumex hydrolapathum	I (3)
Lemna minor	I (3)
Cardamine pratensis	I (1)
Mentha aquatica	I (3)
Carex riparia	I (2)
Number of samples	42
Number of species/sample	8 (3–13)
Vegetation height (cm)	109 (90–200)
Vegetation cover (%)	86 (70–100)

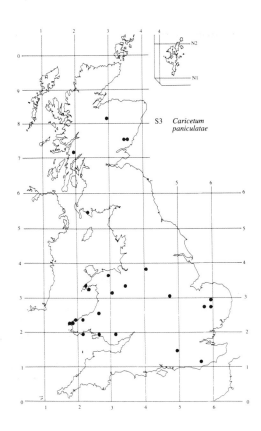

S3 *Caricetum*
    paniculatae

Carex appropinquata in fens

Carex appropinquata is a rare sedge in Britain, occurring mainly in East Anglia with outlying stations in west Dyfed, North Yorkshire and Humberside and the Scottish borders (Jermy *et al.* 1982). Although it can be locally abundant and a dominant of fen vegetation in this country, it does not seem here to constitute the characterising species of a distinct *Caricetum appropinquatae* such as has been recognised on the Continent (e.g. Tüxen 1937, Westhoff & den Held 1969, Oberdorfer 1977).

In Britain, it usually occurs with a combination of species characteristic of rich-fen vegetation, tall herbs such as *Phragmites australis, Calamagrostis canescens, Lysimachia vulgaris, Lythrum salicaria* and *Peucedanum palustre* and other, often bulky components of a subsidiary rush/sedge layer, such as *Juncus subnodulosus, Carex elata, C. diandra* and *C. lasiocarpa*. In this scheme, such vegetation has been classified in a number of communities in which *C. appropinquata* is considered as a local dominant: the *Peucedano-Phragmitetum, Cicuta* sub-community, the *Potentillo-Caricetum rostratae* and the calcicolous *Carex rostrata-Calliergon* mire (see also Wheeler 1980*d*).

S4

Phragmites australis swamp and reed-beds
Phragmitetum australis   (Gams 1927) Schmale 1939

Synonymy

Scirpo-Phragmitetum   Koch 1926 *p.p.*; *Scirpeto-Phrag-mitetum medioeuropaeum*   (Koch 1926) R.Tx. 1941 *p.p.*

Constant species

Phragmites australis.

Rare species

Cicuta virosa, Utricularia intermedia.

Physiognomy

All the vegetation types included here are characterised by the generally overwhelming dominance of *Phragmites australis*. However, this is a very polymorphic species and the gross appearance of the vegetation, even with pure stands, can be very variable. Much of this variation is known to be phenotypic adaptation perpetuated in clonal populations and variously described as 'kinds', ecotypes, biotypes or biotopes of reed (e.g. Rudescu *et al*. 1965; Björk 1967; Haslam 1971*a*, 1972*a*, Dykyjová 1978*a*); some is related to genotypic differences (e.g. Björk 1963, 1967; van der Toorn 1972; Riacu *et al*. 1972; Pazourková 1973: Dykyjová 1978*a*). Although variation in Britain is said to be less than elsewhere (Haslam 1972*a*) and presumed to be phenotypic, there has been no systematic study of its extent in this country or of its relationships to the kind of environmental differences that are reflected in the floristics of the various kinds of *Phragmitetum*. Here, it is possible only to give a very general indication of the reed morphology in each of the sub-communities.

Phragmites is normally a highly gregarious species and individual stands of the community can be very extensive. The vegetation is generally very species-poor and no other species attains even occasional frequency throughout. The *Phragmitetum* is, however, very variable and individual stands may show marked peculiarities of composition.

Sub-communities

***Phragmites australis* sub-community:** *Phragmites communis* reedswamp   Tansley 1911; *Phragmites-Scirpus* associes   Pearsall 1918 *p.p.*; *Phragmites communis-Sparganium minimum* and *Phragmites communis-Littorella* open sociations and *Phragmites communis* society   Spence 1964; *Phragmites* monodominant stands   Haslam 1971*a*; *Phragmites nodum*   Daniels 1978; Association of *Phragmites australis* and *Schoenoplectus lacustris* Pigott & Wilson 1978 *p.p.*; *Phragmites communis* swamp sociation   Wheeler 1978; *Phragmites australis* reedswamp, species-poor variant   Meres Survey 1980; *Scirpo-Phragmitetum typicum*   Wheeler 1980; *Phragmites communis* nodum   Adam 1981 *p.p.*; this sub-community includes pure and very species-poor swamps and reed-beds in which *Phragmites* is the sole constant. The reed cover can be open or closed but *Phragmites* is always the most abundant helophyte forming a canopy from about 1–3 m high. Other species can, however, be locally prominent including other swamp dominants such as *Typha latifolia, T. angustifolia, Carex riparia, C. acuta, Glyceria maxima, Cladium mariscus, Scirpus lacustris* ssp. *lacustris* and *S. maritimus*, tall herbs like *Iris pseudacorus* and *Berula erecta*, sprawlers such as *Solanum dulcamara* and *Calystegia sepium* and aquatic species like *Lemna minor, Callitriche stagnalis, Sparganium minimum* or *Littorella uniflora*. Bryophytes are generally absent.

***Galium palustre* sub-community:** Reed-swamp   West 1905 *p.p.*; *Phragmites communis-Galium palustre* sociation   Spence 1964 *p.p.*; includes *Cicuto-Phragmitetum* Wheeler 1978. Here, the reed cover tends to be a little more open, although it can still be very tall, from 1.5 to 3 m. The vegetation is also somewhat richer with *Galium palustre* constant, *Mentha aquatica* frequent and tall herbs such as *Lythrum salicaria, Iris pseudacorus* and *Epilobium hirsutum* occasional, though none of these species is consistently abundant. There is sometimes a little *Calliergon cuspidatum*, although other bryophytes are rare. Again, a variety of other species can be locally prominent including *Sparganium erectum*, *Carex pseudocyperus* and *Solanum dulcamara*. Some highly distinctive but very locally distributed *P. australis* swamp in which the tall emergent herbs *Cicuta virosa*, *Ranunculus lingua*, *Rumex hydrolapathum* and *Sium latifolium* occur together, with some *Carex pseudocyperus*, *Scirpus lacustris* ssp. *lacustris* and *Typha angustifolia* (the *Scirpo-Phragmitetum* of Wheeler 1978, 1980a; see also the *Ranunculus-Cicuta* variant of Spence 1964) is probably best included in this sub-community.

Menyanthes trifoliata **sub-community:** Reed-swamp Rankin 1911; *Phragmitetum* Matthews 1914; Reed-swamp Holdgate 1955b *p.p.*; *Phragmites communis-Galium palustre* sociation Spence 1964 *p.p.*; *Schoenoplectus lacustris-Phragmites communis* Association, *Phragmites communis-Equisetum fluviatile* subassociation Birks 1973. In these swamps, there is an open reed cover and an understorey of generally small amounts of two of the following: *Menyanthes trifoliata*, *Equisetum fluviatile*, *Carex rostrata* and *Potentilla palustris*. The water surface may have *Nymphaea alba*, *Potamogeton polygonifolius* and *P. natans* and there is sometimes some submerged *Juncus bulbosus* and a variety of aquatics including *Utricularia* spp.

Atriplex prostrata **sub-community:** *Phragmites* marsh Ranwell 1961; Saline/non-saline transition swamp Gimingham 1964; *Phragmitetum* Chapman 1964; *Phragmites communis* reedbeds Proctor 1980; *Phragmites communis* nodum Adam 1981 *p.p.*. The stands included here are dominated by a sometimes open cover of often short *Phragmites*. The most frequent associate throughout is *Atriplex prostrata* but it is not invariably present and, though sometimes abundant, it can be very unevenly distributed in individual stands over local accumulations of organic detritus. *Agrostis stolonifera* may also be present with *A. prostrata* or replace it as the most frequent subsidiary species. Many stands also have halophytes. This somewhat flexible combination of features marks off these kinds of reed-beds from those similarly species-poor types in the *Phragmites* sub-community. Three variants can be recognised.
Atriplex prostrata **variant.** *A. prostrata* is a constant, but often the sole, associate beneath a usually closed canopy of *Phragmites* generally about 1.5 m tall.
Puccinellia maritima **variant.** *A. prostrata* is less frequent here but *Puccinellia maritima* forms a sometimes extensive sward beneath the rather open and generally very short (about 75 cm) reed. A variety of other species characteristic of Puccinellion communities may occur and are sometimes locally abundant: *Aster tripolium*, *Plantago maritima*, *Halimione portulacoides* and *Salicornia dolichostachya*. There is frequently a surface mat of algae.
Agrostis stolonifera **variant:** Brackish water communities Birks 1973 *p.p.*; *Phragmites-Agrostis stolonifera* community Wheeler 1980a. *A. prostrata* remains frequent but the distinctive feature is an open or closed mat of sprawling *A. stolonifera* often with some *Festuca rubra* and *Juncus gerardi* and other upper-marsh halophytes.

Habitat
Phragmites is a natural dominant in these vegetation types in a wide range of permanently wet or periodically waterlogged habitats of differing trophic state and with a variety of substrates. Stands are common in open-water transitions around lakes and ponds, in flood-plain mires and in estuaries, where their extent can be considerable, along dykes (including those with brackish water), canals and sluggish lowland rivers, in small pools, peat cuttings and on salt-marshes. The artificial dominance maintained by cropping for reed extends the occurrence of the community into some naturally drier situations.

The success of *Phragmites* as a dominant in such a diversity of habitats is dependent, to a great extent, on its growth habit (see, for example, Rudescu *et al.* 1965; Haslam 1969a, b, c, 1970a, b, 1971c, 1972a; Rodewald-Rudescu 1974; Fiala 1976, 1978). It is a rhizomatous perennial with annual aerial parts and the normal pattern of growth is as follows. Towards late summer, when underground food reserves are attaining a maximum after downward transfer of material from the shoots, a bud grows out from the base of the previous year's vertical rhizome and extends horizontally for some distance. Its apex turns up and the tip then remains dormant, close to the substrate surface where the developmental transition to a shoot will take place, from November to the following spring. In Britain, the emergence of the young shoots (colts) generally begins between late March and late April, continuing, normally in a single sustained burst, for some 1–3 months so that, by June or July, shoot density is at a maximum. Larger buds tend to sprout first and these grow faster to produce taller shoots. Some late emergent buds may go on to produce shoots through the summer but these may suffer in the competition for light within the stand or be killed by frost before or after their growth is arrested by the cooler temperatures after September. Any shoots which are killed after the main emergence period are not replaced. As the shoots extend, the leaves unfurl and, around the vertical rhizomes and shoot bases, there is an extension of the thick felted mass of roots.

Inflorescences begin to emerge from the aerial shoots between late July and early August and they flower about a month later, the fruit ripening by November and being shed through the winter and spring. As summer progresses, however, nutrients begin to be cycled into the rhizomes, the stems start to harden and the lower leaf blades to drop. Abscission continues into the winter and, by January, most blades have been shed and the stems are dead and brittle. These may remain standing for two or three seasons, as the new annual growth comes and goes, after which they break off close to the ground leaving a stubble which can persist for several more years.

Also in late summer, several new buds begin to form on the upper region of the vertical rhizome. Most of these develop before winter, when they become dormant, and they will produce the aerial growth of the following season, a cluster of shoots, shorter and thinner than the single first generation shoot and less likely to flower. At this time too, as nutrients begin to accumulate again below ground, there is a renewed round of growth in the horizontal rhizome system, new buds growing out from the bases of the vertical rhizomes, extending, branching and developing long, sparse, roots, then turning up ready to initiate new aerial growth in the following spring. After about three seasons, the rhizome system begins to die from behind.

The general appearance of any stand of the *Phragmitetum* is a function of the density and diameter of the buds which exert a primary control on the density and height, respectively, of the aerial shoots (Haslam 1971*b*). These bud characteristics show an inverse relationship with one another and this is perhaps indicative of a potentially equal performance within particular biotopes (Haslam 1971*a*, van der Toorn & Mook 1982). In Britain, the modal density of monodominant stands of *Phragmites* is more than 100 shoots m^{-2} and the modal height is usually over 1 m but apparently stable stands can be found with over 200 shorter shoots m^{-2} or as few as 30 shoots m^{-2} with a modal height of 2.5 m (Haslam 1971*a*). The density and diameter of the developing buds are strongly influenced by the environment of the upper layers of the substrate, especially in late summer and spring, and the extent to which their growth potential is realised is further affected by the environment, both below and above ground, between spring and autumn. The most important variables influencing performance in these ways seem to be the water regime, the trophic state of the medium and the temperature of the environment and their effects are complex and interactive. The water and nutrient regimes also appear to be the major influences on the subsidiary floristic variation visible in the sub-communities.

In general, *Phragmites* performs best and stands of the community are most luxuriant and productive in wet, eutrophic habitats where there is a warm summer. *Phragmites* can survive in a wide variety of water regimes, with water-tables which range between 2 m

above the substrate to more than 1 m below and with various patterns of fluctuation or none (Haslam 1970*c*). The *Phragmitetum* as defined here includes stands from permanently deep and permanently shallow waters, from summer-dry but winter-flooded sites, from places which are permanently moist but never flooded and from situations where there is periodic flooding by fresh, brackish or tidal waters. The maintenance of healthy growth does, however, seem to be favoured by a regularity in the water regime, whether this involves fluctuation or not: sudden changes in water level can severely disrupt the natural rhythm of vegetative growth and may allow vigorous competitors to gain a hold (Haslam 1970*c*, 1971*a*, *b*). The community is noticeably uncommon in habitats which are subject to erratic variation in water level, such as the draw-down zones of reservoirs and spatey rivers.

Phragmites can grow well in the reducing conditions which result from waterlogging provided rhizome aeration is maintained (Haslam 1970*c*). This happens through the dead aerial stems, the natural persistence of which roughly matches the life of the rhizomes. If stems become broken and submerged, as may happen with wave action or where cut stubble is flooded too deeply, aeration is impeded and the bud inception in late summer, and perhaps also that in the spring, is reduced (Rudescu *et al.* 1965, Haslam 1970*c*, 1972*b*). Under normal circumstances, standing high water actually favours the development of those thicker and lower autumn buds which have a maximum height potential (Haslam 1971*a*).

The best performance seems to be attained in Britain where the water level ranges from $+50$ cm to -20 cm and where there is flooding for at least several months of the year (Haslam 1970*c*). The dense stands of tall and vigorous *Phragmites* that can develop under such conditions here fall mainly in the *Phragmites* sub-community which is especially characteristic of open-water transitions and flood-plain mires with permanent standing water or a long winter and spring flood and of deeper and wider dykes which never dry.

In the more nutrient-rich of these deeper waters, *Phragmites* has few competitors. *Scirpus lacustris* ssp. *lacustris* is perhaps more tolerant of exposure to wave action and better able to get a hold on coarser substrates but, of other helophytes, only *Glyceria maxima* and *Typha* spp. seem to be able to challenge *Phragmites* and then only under particular conditions (see below). In general, stands of the *Phragmites* sub-community are extensive and species-poor and other species attain only local and very patchy prominence.

In shallower waters, however, or where flooding is rare and/or brief, the growth of *Phragmites* may be less dense and tall and subsidiary species may gain a hold in the more open shallow water, on the litter which begins to accumulate without prolonged or frequent flooding or on the periodically exposed or accreting substrate, provided they can tolerate a certain amount of shade (Haslam 1971*a*). The *Galium* sub-community comprises such vegetation on floating rafts of peat over loose silts or on firm organic or mineral material in open-water transitions and flood-plain mires and along water courses where there is a water-table that is below the surface for usually eight months of the year and perhaps often below the upper reaches of the vertical rhizomes in late summer when the following year's buds are forming. In Scottish lakes, Spence (1964) noted depth means and ranges of −10 (−40 to +2) cm for the *Galium* sub-community as against +13 (−13 to +50) cm for the *Phragmites* sub-community.

The depth penetration of *Phragmites* in any given waters seems to be limited more by the unavailability of nutrients than by any intolerance of prolonged water-logging (Haslam 1970*c*). Unlike many hydrophytes, the leaves of *Phragmites* die and rot underwater (Hürlimann 1951) and the plants must be capable of sufficiently tall and dense growth to be able to put an effective photo-synthesising canopy above the surface. Roughly speak-ing, about one third or more of the shoot must be out of water (Haslam 1973) and this may not be attainable where short supply of nutrient limits growth. Although *Phragmites* seems to have a small absorption plasticity (Dykyjová 1978*b*), it certainly grows better in more eutrophic conditions. Both nitrogen and phosophorous have been found to be limiting in a variety of situations (Misra 1938, Hürlimann 1951, Björk 1967, Haslam 1965, 1971*c*, van der Toorn 1972) and nutrient defi-ciency can limit bud density (Björk 1967, Haslam 1971*a*), the development of buds even in initially dense clusters (Haslam 1970*b*) and bud diameter and subse-quent aerial growth (Haslam 1971*a*). Nutrients can originate from the waters or from existing or accreting substrates and; provided there is an adequate supply, *Phragmites* seems to show few substrate preferences, growing equally well on wholly organic and wholly mineral material (Haslam 1972*a*). However, since it is a good peat former, it is particularly associated with organic deposits.

In more oligotrophic waters over acid peats or nutri-ent-poor silts, a thinner cover of *Phragmites* may be found in association with swamp species more tolerant of nutrient-poor situations. The *Menyanthes* sub-community is especially distinctive of such conditions around the margins of lakes to the north and west.

Phragmites is moderately tolerant of saline waters and soils: in the Exe estuary in Devon, it occurs over a range of salinities from 2 to 12 g l⁻¹ (Proctor 1980) and in Poole Harbour, Dorset, it survives in salinities up to about 22 g l⁻¹, close to its experimentally determined limit (Ranwell *et al.* 1964, Ranwell 1972). In less saline waters, quite dense growth is possible and the *Phrag-mites* sub-community includes some monodominant emergent stands from brackish dykes and estuaries.

Salt, however, is known to decrease or even prevent bud formation at higher concentrations (Ranwell *et al.* 1964, Haslam 1969*a*, 1972*b*) and to decrease bud dia-meter (Haslam 1971*a*) and stands of the *Phragmitetum* on salt-marshes, though sometimes extensive, often have a rather sparse cover of short shoots. Under such conditions, where there is greater light penetration, there may be a dense subsidiary flora and many stands of the *Atriplex* sub-community are from these situations. The composition of the understorey is somewhat vari-able, being partly dependent on the regime of tidal inundation and partly on the accumulation of litter which is readily trapped among the reed shoots. At some sites, a very open cover of diminutive *Phragmites* occurs over what is essentially a *Puccinellietum* sward (the *Puccinellia* variant) or even with *Zostera noltii* at low levels on salt-marshes. Such stands may have some sub-surface seepage of fresh water which ameliorates the high salinities resulting from frequent inundation. In other cases, there is a *Juncetum gerardi* kind of under-storey at higher levels (some stands in the *Agrostis* variant). Where litter becomes trapped, its decay may release a flush of nitrogen which is often reflected in a patchy cover of *Atriplex prostrata* (*Atriplex* variant). Upper-marsh sites with water seepage may have a thick mat of *Agrostis stolonifera* (some stands in the *Agrostis* variant). Very species-poor salt-marsh stands lacking halophytes are indistinguishable floristically from some *Phragmitetum* in litter-choked or frequently inundated silty sites around inland fresh waters (like those in the *Phragmites-Agrostis stolonifera* community of Wheeler 1980*a*).

Although *Phragmites* has been reported in Britain from up to about 500 m, it is essentially a lowland plant and the *Phragmitetum* is commonest below 150 m. This is partly a reflection of the distribution of more eutro-phic waters but it is probably related to climatic varia-tion and particularly to summer temperatures. *Phrag-mites* becomes more sterile towards its northern European limit (Dahl 1934, Haslam 1972*a*) and in Britain must complete most of its cycle of annual growth between April and September. Warmer temperatures influence the attainment of the height potential in aerial shoots by stimulating intercalary and apical growth (Haslam 1969*c*) but this relationship is confused by the influence of other variables. In Scotland, for example, Spence (1964) calculated a regression of the height of flowering shoots on the mean temperature of the war-mest month, but his equation is unlikely to be of national application because waters in the cooler cli-mates of Scotland tend to be more oligotrophic and acid than elsewhere. *Phragmites* from colder parts of Europe but in richer waters can be taller than Scottish plants (Björk 1967).

A second and more local effect of temperature concerns frosts, especially those in spring. Spring frosting can have a marked effect on the performance of *Phragmites* by resulting in an increase in bud density and a lengthening of the period of emergence. These more numerous buds are smaller than would otherwise have been the case and the aerial shoots consequently shorter (Haslam 1971*a*, 1972*a*). Emerging later, they may also suffer from competition for light (van der Toorn & Mook 1982). Moderate frosts seem to be the most effective in producing such changes; very heavy or repeated frosts can kill all emergent shoots and effectively prevent their replacement in that season. Such damage can result in marked short-term variation in the appearance of stands of the *Phragmitetum* and this may take some seasons to correct itself (Mook & van der Toorn 1982).

The impact of frosting or of other forms of damage to the buds, such as the scorching that can result from intense burning of the vegetation, is mediated through standing water or a litter cover which insulate against marked temperature variations (Haslam 1971*a*). With prolonged or frequent inundation, fallen leaves and stems slowly disintegrate to fine particulate material which may be readily washed away (Mason & Bryant 1975). Where flooding is rare or brief, however, and where the standing crop of dead stems is not harvested, decaying material accumulates to form a thick mat of litter which may contribute up to 40% of the total crop in September (Wheeler & Giller 1982*a*). If this is disturbed by animals or burned off, the buds may be exposed to damage. Haslam (1969*b*, 1970*b*, 1971*a*) proposed a variety of hypotheses to account for different kinds of damage including the killing of existing buds and stimulation of new initials by frosting, and the breaking of dormancy by scorch. Van der Toorn & Mook proposed a simpler model to account for the same effects: that *Phragmites* produces a surplus of autumn buds of which, under normal circumstances, only the dominant apical one undergoes development but in which, if this dominance is removed for whatever reason, all begin to sprout.

When growing well, the *Phragmitetum* is amongst the most productive of all swamp communities, commonly producing a standing crop in Britain of up to 1 kg m^{-2} and sometimes approaching 2 kg m^{-2} (Buttery & Lambert 1965, Haslam 1972*b*, Mason & Bryant 1975, Wheeler & Giller 1982*a*). The distinctive growth habit which leaves an annual crop of dead, foliage-free stems erect, accessible and often with virtually no remains of other plants, makes the vegetation ideal for winter harvesting. Moreover, the removal of this material helps keep down litter and exposes the buds to frost which, if

not too severe or frequent, can thicken up the vegetation and increase the crop. Dense and almost pure stands of the *Phragmitetum* can be maintained under such a regime and natural succession to fen and scrub prevented provided certain conditions are fulfilled. Other species which can readily germinate on the open and almost litter-free cut bed in spring must be destroyed: traditionally this has often been done by natural or artificial flooding. This, however, must not be so deep as to inundate the stubble and impede rhizome aeration, so retarding autumn bud development. And, if heavy or repeated cropping threatens to drain the nutrient reserves of the habitat, flooding with eutrophic waters or artificial fertilising must be undertaken (McDougall 1972, Haslam 1972*b*). The *Phragmites* and *Galium* subcommunities both contain stands of *Phragmites* treated as commercial reed-beds but floristically indistinguishable from natural swamps.

Cutting of green *Phragmites* after the main emergence period, on the other hand, results in a loss of irreplaceable shoots and a depressed yield in subsequent years (Haslam 1972*b*). Indeed, summer mowing of mixed stands of *Phragmites* and *Cladium mariscus* has been a standard way of purifying sedge-beds and, when combined with either deep-flooding or drainage such cutting can help eradicate *Phragmites*.

There is also some loss of reed from stands of the *Phragmitetum* because of grazing and browsing by both stock and wild animals including coypu, deer, water vole (Haslam 1972*a*) and certain wildfowl. Light grazing need not be deleterious because thicker crops of shorter shoots can be produced if leading emergents are bitten off or if disturbance of the litter mat exposes buds to frost (Haslam 1969*a*, 1971*a*). However, heavy grazing after emergence may prevent shoot replacement and trampling can damage the upper rhizomes and hinder bud development in the autumn (Rudescu *et al.* 1965, Haslam 1969*c*, 1971*a*, 1972*b*). Grazing combined with drainage is especially effective in speeding reed decline (Lambert 1948, Bittman 1953, Spence 1964).

Where grazing is by terrestrial mammals at the landward margins of swamps, the limit is often set by flooding which prevents access (Haslam 1971*a*). Coypu and wildfowl, on the other hand, often have their major impact toward the water's edge. Inland populations of larger geese probably have their greatest effect in late spring when succulent shoots are easily accessible and this, of course, is the time when damage can be most severe because such shoots may not be replaced (Haslam 1969*b*). Geese ingest relatively large quantities of food in relation to their body weight (Ogilvie 1978) and in Czechoslovakian swamps *Phragmites* has been shown to provide up to 90% of the late-spring diet of greylag (Hudec 1973). In Broadland, Boorman & Fuller (1981) concluded that feral greylag and Canada geese had probably contributed to the recent general decline of swamp vegetation, much of which was *Phragmitetum*. It is also possible that certain wildfowl have altered the balance between different swamp helophytes: in Czechoslovakia, preferential grazing by greylag can give *Typha angustifolia* a temporary advantage against *Phragmites* (Fiala & Kvet 1971, Boorman & Fuller 1981).

Probably much more important in Broadland, though, has been the impact of feral coypu. Although these animals have a preference for *T. angustifolia* (Gosling 1974), they will readily eat reed, nibbling off emergent shoots in spring, pulling down leaves in summer and digging for roots and rhizomes in winter (Ellis 1963, Gosling 1974). Although much damage remains unobserved (coypu are largely nocturnal (Gosling 1979) and their activity is often masked by turbid water or shifting silt), various studies have suggested that much, if not all, the substantial swamp loss around the Broads in recent years can be attributed to their sometimes large population and substantial appetites (e.g. Anon. 1978, Gosling 1974, 1975a, b, Boorman & Fuller 1981).

Boorman & Fuller (1981) suggested that such damage might have been exacerbated by the general eutrophication of Broadland waters that has occurred recently, perhaps itself due in part to the faeces of coypu and wildfowl, as well as to the more obvious input of nutrients from fertiliser run-off and sewage. The decline of aquatic macrophytes which has followed this change could have deprived coypu and wildfowl of one food source and concentrated their attention on swamp vegetation; increased deposition of silt could also have made some stands more accessible. There is some evidence, too, that eutrophication can lead to decline in *Phragmites* for other reasons: when emergent shoots, for example, are deprived of light and oxygen by a smothering mat of blooming algae (Klötzli 1971, Schröder 1979) or when low sclerenchyma levels in fleshy reed tissues make stems more susceptible to damage by wave action (Klötzi & Grünig 1976).

As well as providing a food source for some herbivores, larger stands of the *Phragmitetum* can offer a valuable breeding or roosting site for a variety of birds (Fuller 1982). Some species, such as reed warbler (the commonest reed-bed breeder), cuckoo (for which reed warblers provide the commonest host in British wetlands), reed bunting, mallard and moorhen, are by no means restricted to the *Phragmitetum* but most of the British populations of bittern, marsh harrier and bearded tit nest exclusively in the community. Stands vary, however, in their quality for birds, perhaps because of differences in the invertebrate populations which form the prey of smaller passerines or of variation in contiguous wet or dry habitats which are needed as feeding ground: even the best swamps are relatively poor

in breeding species. The community also provides important late summer roots for large numbers of swallow, sand martin and starling.

Zonation and succession

The wide ecological amplitude of *Phragmites* makes the *Phragmitetum* one of the commonest components of zonations in open-water transitions and flood-plain mires and means that a wide variety of other swamp and fen communities, with more exacting species, can be found in association with it. The *Phragmitetum* is also widespread in the more compressed or fragmentary sequences associated with dyke and canal margins and it has a scattered distribution within salt-marsh zonations. It also persists, sometimes because of deliberate treatment as a crop, in the often very complex patterns of vegetation associated with disturbed fens in flood-plain and valley mires.

In more natural situations, the community occurs as part of zonations which can, in any particular site, be related most frequently to a gradient of water-level. In extensive open-water transitions it is often the most distal swamp type giving way directly, in deeper unpolluted waters, to floating-leaved or submerged aquatic vegetation. In some cases, stands are fronted by other swamp types: in lakes to the north and west, there is sometimes a belt of the *Scirpetum lacustris* (Pearsall 1918, Spence 1964) and this pattern was probably, at one time, more widespread (as in Broadland: see Pallis 1911, Lambert & Jennings 1951); in more eutrophic waters, the *Typhetum angustifoliae* occurs in this position (Lambert 1951) (Figure 12) or, in more sheltered sites in the rather peculiar conditions along the Yare valley in Norfolk, the *Glycerietum maximae* (Lambert 1946). Much conflated versions of these sequences occur commonly in mesotrophic and eutrophic dykes and canals (e.g. Haslam 1978).

Where there is a substantial depth of water available for colonisation, it is usually the *Phragmites* sub-community that leads such sequences or, in poorer waters especially over peat, the *Menyanthes* sub-community (e.g. Birks 1973). In waters which are more shallow throughout, the *Galium* sub-community is more usual. Very commonly, a gradual reduction in water-level is matched by the *Phragmites* sub-community being replaced inshore by the *Galium* sub-community: this is the common pattern in larger lakes in Scotland and north-western England (e.g. Spence 1964). In these regions too, lake stands of the *Phragmitetum* tend to be less extensive, being often restricted to more nutrient-rich areas of deposition, as around the deltas of input streams. This means that the community often forms a complex patchwork around the shores with other swamps like the *Caricetum rostratae* and the *Caricetum vesicariae*.

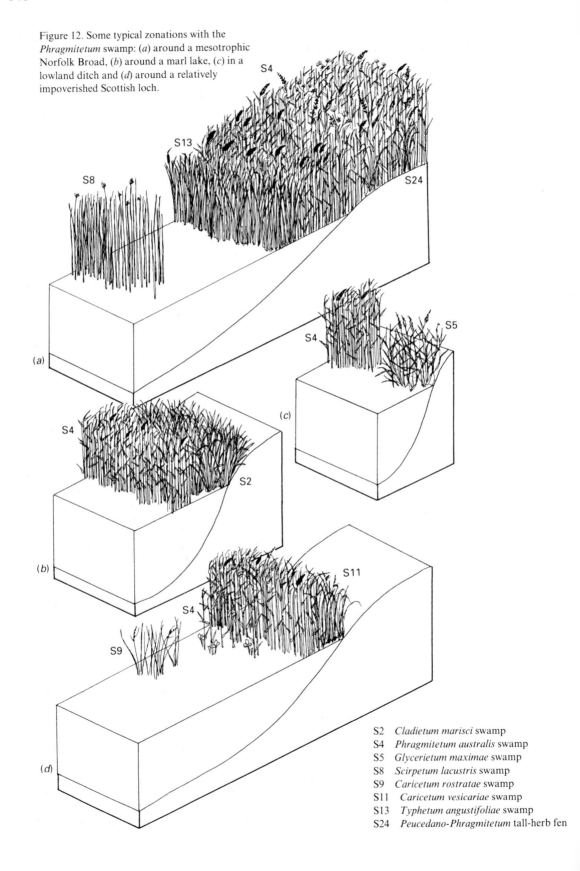

Figure 12. Some typical zonations with the *Phragmitetum* swamp: (*a*) around a mesotrophic Norfolk Broad, (*b*) around a marl lake, (*c*) in a lowland ditch and (*d*) around a relatively impoverished Scottish loch.

S2 *Cladietum marisci* swamp
S4 *Phragmitetum australis* swamp
S5 *Glycerietum maximae* swamp
S8 *Scirpetum lacustris* swamp
S9 *Caricetum rostratae* swamp
S11 *Caricetum vesicariae* swamp
S13 *Typhetum angustifoliae* swamp
S24 *Peucedano-Phragmitetum* tall-herb fen

At some sites, the *Phragmitetum* may give way inshore, but still in standing water, to other swamp types. In oligotrophic calcareous waters, the *Cladietum marisci* may occur here, or in more mesotrophic situations, the *Caricetum paniculatae* (Lambert & Jennings 1951, Lambert 1951). Along the Yare, a narrow front of the *Phragmitetum*, resistant to scour, can protect a bank of the *Glycerietum maximae* (Lambert 1946).

Frequently, however, the community gives way directly to some form of fen. The more landward parts of these zonations are strongly influenced by the trophic state of the substrate and also by human interference which, pushing out to the limit of standing water, has made intact sequences increasingly rare. In the more complete transitions, the *Phragmitetum* may pass gradually, through the *Galium* sub-community and with a progressive increase in the number and variety of associates, to fen vegetation in which *Phragmites* remains a prominent component. On mesotrophic silts and organic soils, there can be a zonations to the *Phragmites-Eupatorium* fen; this is a common, though often rather fragmentary, pattern in some valley mires (e.g. Haslam 1965). On the fen peats of Broadland, the *Phragmitetum* can give way to the very rich and varied *Peucedano-Phragmitetum* (e.g. Pallis 1911, Lambert & Jennings 1951, Wheeler 1978). In more acid and oligotrophic waters, the community may pass to a swinging mat of the *Potentillo-Caricetum rostratae* (e.g. Pearsall 1918, Holdgate 1955b, Pigott & Wilson 1978) which sometimes seems to be able to grow through and beyond the *Phragmitetum*.

In the agricultural lowlands, many transitions of this kind have been truncated by systematic drainage and, even in those few places, like Broadland, where extensive tracts of swamp and fen remain, zonations are very complex and sometimes abrupt because of differences in treatment applied piecemeal to marsh parcels (e.g. Lambert 1948). A very common feature of disturbed and eutrophicated sites throughout the lowlands is the juxtaposition of the *Phragmites* or *Galium* sub-communities with various kinds of *Phragmites-Urtica* fen, the *Phalaridetum* or tall-herb vegetation like the *Epilobietum hirsutae*. Narrow strips of these vegetation types also occur widely in sequences along the banks of dykes.

Although stands of the *Phragmites* sub-community are also found in brackish dykes behind reclamation banks, the majority of salt-marsh and estuarine occurrences of the community are in sites without permanent standing water. Sometimes stands of the *Atriplex* sub-community seem to be simply superimposed upon the existing salt-marsh zonation, occurring within *Puccinellietum* or *Juncetum gerardi* swards. In other cases, this sub-community forms part of the complex of vegetation types that occurs around the upper-marsh transition where there is litter accumulation and/or freshwater seepage. In estuaries, stands of the *Phragmitetum* sometimes occur in the inverted zonations associated with reversals of salinity gradient.

When undisturbed, the *Phragmitetum* can be a very persistent community. *Phragmites* can retain its dominance under a wide variety of conditions and in optimal habitats stands can be very extensive and long-lived: Rudescu *et al.* (1965) estimated the age of some clones as in excess of 1000 years. Although seed-set, fertility and viability can be good (e.g. Hürlimann 1951, Bittman 1953, Spence 1964, Haslam 1972a), conditions are rarely favourable for the growth of seedlings within existing stands, but this may be of little consequence with such an efficient system of vegetative renewal. *Phragmites* can attain a very high shoot density on healthy advancing fronts (Haslam 1971a, 1972a) and retain it within older stands, producing a canopy that, by mid-summer, has reduced light penetration below very substantially (Buttery & Lambert 1965, Haslam 1971a, 1972a). In deeper water, it has few natural competitors and, on drier ground, an undisturbed litter mat can help prevent the establishment of other species (Haslam 1971a). Stands which suffer moderate damage from frosting, burning or insect infestation seem to possess an ability to restore a balanced shoot production within a few seasons (Mook & van der Toorn 1982).

The essential feature of successful competitors to *Phragmites* seems to be that they can capitalise on light levels under a given water and nutrient regime. Some achieve this by virtue of a different phenology to reed. Both those swamp helophytes which seem able to compete with *Phragmites* in deeper waters, *Glyceria maxima* and *Typha angustifolia*, attain maximum shoot densities and a critical light interception by May, when the emergence of reed shoots is only one third or so complete (Buttery & Lambert 1965, Mason & Bryant 1975). Moreover, *G. maxima* goes on producing shoots until late in the season and not until the shoots lodge is light penetration increased.

Some other species require a drier substrate to invade. *Cladium mariscus* and *Carex paniculata* both seem able to take advantage of the period between the emergence of an accumulating *Phragmites* root felt above the water level and the build-up of litter on the now drier surface (e.g. Lambert 1951). Moreover, both are evergreen and somewhat shade-tolerant and well adapted to survive the gradual depression of the mat of *Phragmites* under the water with the slowly increasing weight of their bulky stocks. *C. mariscus* may have the additional advantage of a greater tolerance of oligotrophic conditions than *Phragmites* (Lambert 1951).

The establishment of most other species within the *Phragmitetum* depends upon some reduction in reed vigour to open the canopy and an amelioration of the blanketing effect of the accumulating litter mat. There may be an opportunity for some species to invade in those situations where, with slow terrestrialisation, a still thin litter mat is exposed to decreasing periods of winter submergence with ever shallower waters. The development of the *Galium* sub-community from the *Phragmites* sub-community occurs in this way as relatively shade-tolerant herbs are able to establish on the damp litter in spring. As the reed cover becomes less vigorous on the drier substrate, there may be a continuing natural succession to the richer vegetation of the *Phragmites-Eupatorium* or *Peucedano-Phragmitetum* fens and thence to fen carr (e.g. Tansley 1939, Lambert & Jennings 1951, Lambert 1951, Wheeler 1980c). At various stages in such sequences, winter-cropping of reed can help the *Phragmitetum* persist against the tendency towards invasion of competitors or confuse the natural succession.

More drastic disturbance of the water regime or of the litter mat may produce a strong disruption in the natural growth rhythm of *Phragmites*, giving any invaders a season or more's advantage. The rapid drop in water level consequent upon drainage can produce such results (Haslam 1970c, 1971a, b) as can very severe and repeated frost or mismanagement of reed-beds. If there is surface disturbance, eutrophication of oxidising organic matter is often followed by the development of the *Phragmites-Urtica* community with its very characteristic patches of nitrophilous tall herbs (e.g. Haslam 1965). The enrichment of dyke waters by fertiliser run-off and sewage and the disturbance involved in dyke clearance and dredging is often marked by successions to this fen on the banks.

On some ungrazed salt-marshes, the *Phragmitetum* has replaced parts of stands of the *Spartinetum townsendii* but the status of the community in general successions of salt-marshes is difficult to assess (Ranwell 1961, 1964a, 1972).

Distribution
The *Phragmites* sub-community is widespread throughout the British lowlands. The *Galium* sub-community

has also been encountered in most regions, although it appears especially conspicuous around Scottish lakes where it is the major drier *Phragmites*-dominated community, replacing such fens as the *Phragmites-Eupatorium* community and the *Peucedano-Phragmitetum*. The *Menyanthes* sub-community has been recorded only from isolated localities in the north-west and Scotland.

The variants of the *Atriplex* sub-community have a more patchy distribution on coastal and inland salt-marshes and (without halophytes) around some inland freshwater bodies. Although the *Phragmitetum* is rare on the western Scottish coast (Adam 1978), some of the largest intertidal stands in the country occur in eastern Scotland, notably on the Tay (Ingram *et al.* 1980).

Affinities
The *Phragmitetum* as diagnosed here is a narrower unit than some of the reed-swamp communities defined from mainland Europe, such as the *Scirpo-Phragmitetum* of Koch (1926) or its amendments like the *Scirpeto-Phragmitetum* Tüxen 1941 which have included vegetation dominated by other swamp helophytes. As with all species-poor swamps, there may be difficulties of definition where *Phragmites* dominance is giving way to species such as *Scirpus lacustris* ssp. *lacustris*, *Typha angustifolia*, *Cladium mariscus* and *Carex paniculata*. Mixed stands where *Phragmites* has a quantitatively subordinate role have been assigned to sub-communities in the vegetation types dominated by these other species.

More species-rich stands of the *Phragmitetum* show affinities to the often *Phragmites*-dominated vegetation of the *Phragmites-Eupatorium* fen but the more consistent presence there of species such as *Lythrum salicaria*, *Eupatorium cannabinum*, *Filipendula ulmaria* and *Iris pseudacorus* is generally a good demarcation. Likewise, although stands of the *Phragmitetum* may have scattered clumps of tall herbs such as *Urtica dioica* and *Epilobium hirsutum*, these are much more frequently conspicuous in the *Phragmites-Urtica* fen.

Floristic table S4a

	4a
Phragmites australis	V (5–10)
Solanum dulcamara	I (1–5)
Typha latifolia	I (1–6)
Lemna minor	I (4–8)
Carex riparia	I (1–3)
Iris pseudacorus	I (3–5)
Typha angustifolia	I (2–5)
Carex acuta	I (2–4)
Glyceria maxima	I (1–4)
Oenanthe crocata	I (1–4)
Callitriche stagnalis	I (2–5)
Cladium mariscus	I (1–5)
Calystegia sepium	I (1–8)
Carex lasiocarpa	I (5–7)
Berula erecta	I (1–5)
Scirpus maritimus	I (2–5)
Carex panicea	I (3–7)
Ranunculus flammula	I (2–3)
Cardamine pratensis	I (2–3)
Hydrocotyle vulgaris	I (3–4)
Juncus effusus	I (1–3)
Salix cinerea sapling	I (1–3)
Sphagnum recurvum	I (5–7)
Molinia caerulea	I (1–10)
Eriophorum angustifolium	I (4)
Galium aparine	I (2)
Rubus fruticosus agg.	I (1)
Number of samples	176
Number of species/sample	3 (1–12)
Vegetation height (cm)	182 (70–300)
Vegetation cover (%)	94 (20–100)

Floristic table S4b

	4b
Phragmites australis	V (5–10)
Galium palustre	IV (1–5)
Mentha aquatica	III (1–6)
Epilobium hirsutum	II (1–4)
Iris pseudacorus	II (1–5)
Lythrum salicaria	II (1–4)
Calliergon cuspidatum	II (1–4)
Sparganium erectum	I (3–7)
Carex pseudocyperus	I (1–8)
Sium latifolium	I (1–3)
Rumex hydrolapathum	I (1–3)
Carex riparia	I (1–4)
Typha angustifolia	I (1–3)
Lemna minor	I (2–4)
Berula erecta	I (1–3)
Cicuta virosa	I (1–3)
Juncus subnodulosus	I (1–3)
Lotus uliginosus	I (2–3)
Angelica sylvestris	I (1–3)
Solanum dulcamara	I (1–4)
Lychnis flos-cuculi	I (2–3)
Equisetum palustre	I (1–4)
Lycopus europaeus	I (2–3)
Phalaris arundinacea	I (1–3)
Urtica dioica	I (1–2)
Poa trivialis	I (3–4)
Holcus lanatus	I (3–4)
Agrostis stolonifera	I (1–5)
Filipendula ulmaria	I (6)
Ranunculus lingua	I (1)
Brachythecium rutabulum	I (2)
Cardamine pratensis	I (3)
Equisetum fluviatile	I (4)
Myosotis laxa caespitosa	I (3)
Myosotis scorpioides	I (1)
Oenanthe fistulosa	I (1)
Salix cinerea sapling	I (1)
Number of samples	61
Number of species/sample	8 (2–15)
Vegetation height (cm)	205 (150–300)
Vegetation cover (%)	99 (95–100)

Floristic table S4c

	4c
Phragmites australis	V (5–10)
Menyanthes trifoliata	IV (1–4)
Equisetum fluviatile	IV (1–3)
Carex rostrata	IV (2–4)
Nymphaea alba	III (1–6)
Potentilla palustris	III (1–3)
Potamogeton polygonifolius	II (1–3)
Juncus bulbosus	II (2–4)
Potamogeton natans	II (3)
Utricularia minor	I (2)
Eleocharis palustris	I (2)
Juncus articulatus	I (3)
Potamogeton gramineus	I (1)
Carex lasiocarpa	I (2)
Eriophorum angustifolium	I (3)
Sparganium minimum	I (1)
Utricularia intermedia	I (1)
Utricularia vulgaris	I (1)
Hippuris vulgaris	I (1)
Number of samples	8
Number of species/sample	6 (3–10)
Vegetation height (cm)	150
Vegetation cover (%)	78 (20–100)

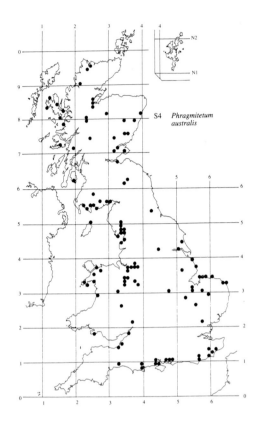

S4 *Phragmitetum australis*

Floristic table S4d

	4di	4dii	4diii
Phragmites australis	V (9–10)	V (6–10)	V (5–10)
Atriplex prostrata	V (1–6)	III (6–10)	III (1–5)
Puccinellia maritima		V (2–7)	II (2–3)
Aster tripolium	I (1)	III (2–4)	
Algal mat	I (5–6)	III (5–8)	I (6–7)
Plantago maritima		II (3–6)	I (2–4)
Salicornia dolichostachya		II (2–3)	I (2)
Halimione portulacoides		II (1–9)	
Cochlearia anglica		I (2–4)	
Limonium cf. *vulgare*		I (1–3)	
Agrostis stolonifera	I (1)	I (4)	V (3–7)
Juncus gerardi		I (3–4)	III (2–8)
Festuca rubra	I (3–4)	I (4–8)	II (2–6)
Armeria maritima		I (4)	I (2–4)
Amblystegium riparium			I (2–6)
Oenanthe lachenalii			I (2–3)
Matricaria maritima			I (1–4)
Carex otrubae			I (2–3)
Calystegia sepium			I (2–6)
Ranunculus sceleratus			I (2–5)
Glyceria fluitans			I (3–5)
Apium graveolens			I (3–4)
Juncus maritimus			I (4–6)
Elymus repens			I (4–6)
Oenanthe crocata			I (2–5)
Juncus articulatus			I (2–3)
Holcus lanatus			I (2–4)
Galium uliginosum			I (2–3)
Rumex crispus			I (2–4)
Sonchus arvensis			I (3–4)
Atriplex littoralis			I (2–4)
Galium aparine			I (3)
Eleocharis palustris			I (3)
Glaux maritima	I (2)	II (2–3)	II (2–8)
Triglochin maritima		II (2–5)	II (2–3)
Scirpus maritimus		II (2–7)	II (1–5)
Elymus pycnanthus	I (3)	I (3–4)	I (5)
Aster tripolium (rayed)	I (2)	I (3–4)	I (1–4)
Cochlearia officinalis		I (3)	I (2–3)
Suaeda maritima		I (3–4)	I (3)
Number of samples	11	21	27
Number of species/sample	3 (2–4)	6 (4–8)	7 (4–17)
Vegetation height (cm)	151 (120–200)	73 (20–150)	147 (40–300)
Vegetation cover (%)	100	91 (40–100)	96 (70–100)

S5
Glyceria maxima swamp
Glycerietum maximae   (Nowinski 1928) Hueck 1931 *emend*. Krausch 1965

Synonymy

Glycerietum aquaticae Tansley 1911; Association à *Scirpus lacustris* et *Glyceria aquatica* Allorge 1922 *p.p.*; *Scirpeto-Phragmitetum glyceriosum aquaticae* Koch 1926; *Glycerietum maximae* and *Glyceria* reedswamp Tansley 1939; *Scirpeto-Phragmitetum medioeuropaeum* (Koch 1926) R.Tx. 1941 *p.p.*; *Glyceria maxima* communities Lambert 1947c *p.p.*; *Glycerieto-Typhetum latifoliae* Neuhäusl 1959 *p.p.*; Societie van *Glyceria maxima* Westhoff & den Held 1969; *Glyceria maxima* sociation Wheeler 1975a *p.p.*

Constant species

Glyceria maxima.

Physiognomy

The *Glycerietum maximae* is always overwhelmingly dominated by *Glyceria maxima* which forms a typically dense and luxuriant cover of leafy shoots, often more than 1 m long and sometimes attaining almost 2 m. The gross appearance of the vegetation is somewhat variable: the *G. maxima* plants may be firmly anchored and the shoots largely erect forming a tall emergent swamp; in other cases, stands occur as swinging masses of marginal 'hover', loosely attached below and with the shoots showing a marked tendency to lodge. Sometimes, stretches of this kind of *Glycerietum* may become detached to form free-floating islands of vegetation. Whatever its physiognomy, the community is typically very species-poor and pure stands are common.

Growth of the leafy aerial shoots of *G. maxima* begins early in the year and stands retain their characteristic bright, fresh green colour throughout the growing season as a succession of new shoots is produced (Lambert 1947c, Buttery & Lambert 1965, Westlake 1966). However, the erect shoots die back rapidly around November to produce a substantial mass of compact, black and slimy litter and so, unlike some swamp communities where the dominants are largely evergreen or where there is a large amount of standing dead material, the winter appearance of the vegetation here is very different from that of the growing season (Tansley 1939, Lambert 1947c, Westlake 1966). Decumbent shoots may remain winter-green, perhaps because of protection from frost or wind (Lambert 1947c).

Sub-communities

***Glyceria maxima* sub-community:** *Glyceria aquatica* reedswamp Pallis 1911; *Glyceria maxima* floating reedswamp Lambert 1946; *Glyceria* society Spence 1964; Tall grass washlands Ratcliffe 1977 *p.p.*. This sub-community includes pure and species-poor stands in which *G. maxima* typically forms a very dense cover. Stands may be extensive and, as well as swamp or washland vegetation with erect plants, marginal and free-floating masses of 'hover' with many decumbent shoots are included here. There are no frequent associates but a variety of other species occurs occasionally and some of these can be locally abundant. *Lemna minor* and *Nymphaea alba* may be prominent on open water between the *G. maxima* shoots and there are sometimes sprawling masses of *Solanum dulcamara* or *Galium palustre*. Towards the south-east, *Berula erecta* and, especially, *Rumex hydrolapathum* can be conspicuous.

***Alisma plantago-aquatica-Sparganium erectum* sub-community:** *Glyceria aquatica* riparian reedswamp Tansley 1911 *p.p.*; *Glycerietum maximae* riparian reedswamp Butcher 1933; *Glyceria maxima* riparian zone and *Glyceria maxima* stagnant reedswamp *p.p.*. Lambert 1947c. Here, *G. maxima* forms an often more open cover in frequently narrow and fragmentary stands. The plants are usually tall and erect but decumbent shoots may trail out into open water. Again, the vegetation is typically species-poor but there is a little more consistent variety here than in the *Glyceria* sub-community. Especially distinctive is the frequent presence of one or more of *Alisma plantago-aquatica*, *Sparganium erectum* and *Rorippa nasturtium-aquaticum* forming a patchy understorey or fringe fronting the *G. maxima* in slightly deeper water. Other occasional species are those characteristic of riparian water margins.

Habitat

G. maxima has high mineral requirements (Lambert 1947c, Petersen 1949, Lang 1967) and the *Glycerietum* is very much a swamp of eutrophic water margins. It is especially characteristic of nutrient-rich, circumneutral to basic mineral substrates, such as certain alluvia, and it can maintain itself on such material even where the waters are stagnant. It also occurs on oligotrophic, neutral or fen peats provided the mineral supply is continually renewed by the movement of richer waters (Lambert 1946, 1947c). The typically lowland distribution of the *Glycerietum* is probably a reflection, at least in part, of these edaphic needs of the dominant rather than being a direct response to climate (Lambert 1947c). The community occurs, usually below 150 m, as a fringe to sluggish rivers and streams, along dykes and canals, in open-water transitions around ponds, lakes and abandoned industrial water bodies and, often more extensively, on regularly inundated flood-plain washlands and in those parts of Broadland where there is a tidally influenced movement of ponded-back fresh water.

Provided eutrophic conditions are maintained, the community seems tolerant of a wide range of water depth. Although both vegetative and reproductive development in *G. maxima* are best favoured by a water-table that is at about substrate level (Lambert 1947c), it can occur as an emergent rooted in up to 80 cm of water and very vigorous stands of the community can persist in stagnant waters that remain permanently above ground. Although *G. maxima* has roots which are markedly less aerenchymatous than those of *Phragmites australis* (Iverson 1949), there is little evidence to suggest that the former suffers greatly under anaerobic conditions (Buttery & Lambert 1965, Haslam 1971b).

The physiognomy of the community does, however, seem to be strongly influenced by the physical properties of the substrate (Lambert 1946, 1947a, c). Stands with erect shoots generally occur on firmer material. Marginal 'hover', on the other hand, develops where the *G. maxima* extends out into deeper water over loose ooze. Here, the buoyancy of the bulky but aerenchymatous aerial shoots tends to pull the roots and rhizomes free of the substrate so that they trail below a floating mass of shoots which, without substantial basal support, tend to topple over.

The ability of such marginal *Glycerietum* to rise and fall undamaged with frequent changes of water-level makes the community ideally suited to take advantage of tidally influenced water bodies such as the broads of the Yare valley in Norfolk. Here, there is a regular

fluctuation in water-level of some 20–30 cm with occasional much greater variation (Lambert 1946), a movement which itself contributes to the loosening of the substrate and marginal vegetation. In such situations, *G. maxima* probably gains some competitive advantage over other swamp species which, while also being able to capitalise on the continual circulation of nutrient-laden waters, have a more rigid growth habit and/or shallow rooting system and which cannot persist as a floating mat.

Rapid lateral water movement, on the other hand, seems inimical to the development of the community. In and around Surlingham Broad on the Yare, Lambert (1946) noted that, wherever there was appreciable tidal scour, the *Glycerietum* was replaced by *P. australis* swamp. Big floods on the Yare also frequently tear away stretches of marginal *Glycerietum* which can persist for some time as free-floating islands before disintegrating or lodging in some sheltered spot (Lambert 1946, 1947c).

In the moving waters of streams and rivers, Butcher (1933) observed that the community was well developed only where the current velocity was less than about 0.5 km h^{-1}. In such situations, the *Glycerietum* generally occurs as the *Alisma-Sparganium* sub-community and the more fragmentary and varied nature of this vegetation is probably due in part to the continual disturbance of the habitat and the ready dispersal and germination of water-borne propagules. Here, too, substrate profiles are often steeply shelving so that zonations in relation to water depth tend to be telescoped into a narrow and rather jumbled marginal zone (Lambert 1947c).

G. maxima is a highly productive and palatable grass which was once much prized as a fodder and litter crop on the Fen washlands and around the Broads (Camden 1586, Curtis 1777, Lambert 1947c, Godwin 1978). In favourable conditions, dense stands of the *Glycerietum* can yield 0.5–1.5 kg ha^{-1} yr^{-1} dry weight (Buttery & Lambert 1965, Westlake 1966). The early start to growth and late lignification of *G. maxima* give a long possible cropping season and the ready ability of the species to tiller and production of repeated shoot generations assist in a quick recovery after mowing. In days of cheap labour, as many as three cuts were taken in a season, early crops being used for fodder and later ones for litter. Subsidary uses were for poor quality thatch, packing material and binding in brick making (Lambert 1947c, Godwin 1978).

Although *G. maxima* provides a good bite for cattle especially when it is young, swamp stands of the *Glycerietum* are generally inaccessible to grazing stock. Wildfowl will eat *G. maxima* where it abuts on to open water, although dense stands of the community provide an uncongenial breeding habitat for even the commonest water-birds (Fuller 1982). Although seed of many kinds is an important component of the diet of wildfowl, seed production in *G. maxima* seems to be generally low (Lambert 1947c). Coypu also eat *G. maxima* and the general decline of reed-swamp in the Broads (Boorman & Fuller 1981) includes some loss of *Glycerietum* by coypu grazing along the Yare.

Zonation and succession

Extensive, virtually pure stands of the *Glyceria* sub-community are quite common and sometimes completely choke small ponds, ox-bows or flat, narrow channels with standing or slow-moving water (Lambert 1947c, Haslam 1978). Frequently, however, the *Glycerietum* occurs as a definite zone in open-water transitions and riparian sequences, typically passing to the *Sparganietum erecti* in deeper water and, above, to the *Phalaridetum aruncinaceae*. Where banks shelve gently, the community may occur in both forms, the *Alisma-Sparganium* sub-community forming a fragmentary transition to the *Sparganietum erecti* and giving way above to a belt of the *Glyceria* sub-community. Such zonations are a very characteristic feature of pools and streams in the English lowlands.

In the rather special conditions along the Yare, the *Glyceria* sub-community often abuts directly on to the open water of the broads as marginal 'hover' or, especially where there is tidal scour, occurs behind a marginal fringe of *Phragmitetum australis*. Away from the open water, the *Glycerietum* usually gives way to some kind of fen vegetation in which both *G. maxima* and *P. australis* are important components, e.g. the *Glyceria* sub-community of the *Peucedano-Phragmitetum* or the *Epilobium hirsutum* sub-community of the *Phragmites-Urtica* fen. Although the abundance and vigour of *G. maxima* in such sequences clearly decline with increased distance from the moving broads water (Lambert 1946, Buttery & Lambert 1965), there does not seem to be a single, simple edaphic gradient related to this change.

There seems no doubt, however, that once established under favourable conditions, *G. maxima* can play a major part in successions from open water to fen. It is capable of rapid growth (Syme 1872, Lambert 1947c), and can readily encroach upon open water provided there is some contact between the vegetation mat and underlying sediments: at the very front of marginal 'hover' where the roots hang free in the water, there is a small reduction in growth and detached islands of *Glycerietum* quickly become chlorotic if they do not make renewed contact with the substrate (Buttery & Lambert 1965).

In sluggish streams, in dykes and in pools, *G. maxima* litter may accumulate in stagnant conditions and aid terrestrialisation. On the Yare, the *Glycerietum* often initiates succession in more sheltered but irrigated situations (Lambert 1946). Here *G. maxima* seems to have a competitive edge over *P. australis*. These species have broadly similar upper limits of growth in relation to water level and both are tolerant, though by virtue of different growth habits, of the tidal rise and fall. *G. maxima*, however, begins growth earlier from shoots which in the previous autumn have already extended up to substrate level (Lambert 1947c) and the expanding leafy shoots rapidly reduce light penetration below (Buttery & Lambert 1965). There is thus a dense cover of vegetation even before the shoots of *P. australis*, with its deep-seated rhizomes and perennating buds, break the surface. Where mixed stands are mown early, *G. maxima*, with its good growth of aftermath, has the additional advantage of quick recovery over *P. australis* which, though starting later, produces a single flush of shoots and attains its standing crop maximum earlier.

Where conditions begin to isolate growing *Glycerietum* on the Yare from the irrigating waters, its productivity has been shown to decline (Buttery & Lambert 1965). This isolation may happen by continued extension of the community into open water with slow peat accumulation towards the fen hinterland or within the relic arms of broads and in dykes which become blocked (Lambert 1946). It has been shown that, as *G. maxima* becomes, for some reason, unable to utilise available major nutrients, which may not themselves be externally limiting, *P. australis* can gain the ascendant because of some greater ability to tolerate conditions generally unfavourable to both species (Buttery & Lambert 1965). Like the other Broadland seres, the successions involving the *Glycerietum* are more complex than at first sight.

Alongside streams and rivers, the accumulation of substrate beneath the *Glycerietum* seems more dependent upon silting than on the build-up of litter (cf. Tansley 1939). Winter floods readily wash away the *G. maxima* remains (Lambert 1947c) and the repeated disturbance of the mineral material itself may continually set back any successional advance of the community which maintains itself as a permanent fringe.

Distribution

The *Glycerietum* is a lowland community, being commonest on the softer rocks of the Midlands and to the east. In Scotland, its major localities are all in kettle-hole lakes in glacio-fluvial clays (Spence 1964) and it is of very restricted occurrence in Wales (Ratcliffe & Hattey 1982).

Affinities

Species-poor *G. maxima* stands have sometimes been described as a form of the more general swamp community *Scirpeto-Phragmitetum* (e.g. Koch 1926,

Tüxen 1937, LeBrun *et al.* 1949). In other schemes, a distinct *Glycerietum* has been recognised, although this has often been defined rather broadly (e.g. Hueck 1931, Jonas 1933 and, in the British descriptive literature, Tansley 1939 and Lambert 1947*c*). Such definitions have usually included more species-rich fen vegetation with prominent *G. maxima* such as that which is here placed within the *Peucedano-Phragmitetum* and the *Phragmites-Urtica* and *Phragmites-Eupatorium* communities. *G. maxima* may be locally dominant in these vegetation types though, in Britain, over a much more restricted geographical range than that occupied by the *G. maxima* swamps. Narrower interpretations of the *Glycerietum* similar to that adopted here have been pursued by Braun-Blanquet & Tüxen (1952), Jeschke (1959) and Passarge (1964). Certain workers (e.g. Westhoff & den Held 1969, Wheeler 1975, 1980*a*) have preferred to recognise a *G. maxima* society to take account of the difficulties of classifying such species-poor vegetation, though, in some cases, such communities have again included some richer stands.

Floristic table S5

	a	b	5
Glyceria maxima	V (5–10)	V (5–10)	V (5–10)
Rumex hydrolapathum	II (2–6)	I (3)	I (2–6)
Lemna minor	II (1–9)		I (1–9)
Berula erecta	I (2–5)		I (2–5)
Solanum dulcamara	I (2–7)		I (2–7)
Nymphaea alba	I (3–4)		I (3–4)
Rumex crispus	I (2–3)		I (2–3)
Mentha aquatica	I (1–4)		I (1–4)
Galium palustre	I (1–6)		I (1–6)
Juncus effusus	I (1–6)		I (1–6)
Riccia fluitans	I (5)		I (5)
Epilobium hirsutum	I (4)		I (4)
Apium inundatum	I (5)		I (5)
Urtica dioica	I (4)		I (4)
Galium aparine	I (3)		I (3)
Filipendula ulmaria	I (4)		I (4)
Iris pseudacorus	I (1)		I (1)
Ranunculus ficaria	I (5)		I (5)
Equisetum palustre	I (1)		I (1)
Lycopus europaeus	I (3)		I (3)
Poa trivialis	I (3)		I (3)
Angelica sylvestris	I (3)		I (3)
Carex riparia	I (1)		I (1)
Alisma plantago-aquatica		III (2–4)	I (2–4)
Sparganium erectum		III (4–6)	I (4–6)
Nasturtium officinale		III (2–4)	I (2–4)
Lythrum salicaria		I (3–4)	I (3–4)
Ranunculus repens		I (1–2)	I (1–2)
Ranunculus acris		I (2–3)	I (2–3)
Carex otrubae		I (4)	I (4)
Myosotis scorpioides		I (3)	I (3)
Phalaris arundinacea		I (6)	I (6)
Cirsium arvense		I (3)	I (3)
Glechoma hederacea		I (2)	I (2)

Floristic table S5 (*cont.*)

	a	b	5
Eupatorium cannabinum		I (4)	I (4)
Calystegia sepium		I (2)	I (2)
Cirsium palustre		I (3)	I (3)
Carex acuta		I (4)	I (4)
Equisetum fluviatile		I (2)	I (2)
Alopecurus pratensis		I (4)	I (4)
Callitriche stagnalis	I (2)	I (3)	I (2–3)
Phragmites australis	I (1)	I (3–4)	I (1–4)
Polygonum amphibium	I (2)	I (4–6)	I (2–6)
Rumex obtusifolius	I (2)	I (2–3)	I (2–3)
Number of samples	104	16	120
Number of species/sample	4 (1–11)	5 (2–8)	4 (1–11)
Vegetation height (cm)	104 (40–180)	94 (40–180)	103 (40–180)
Vegetation cover (%)	94 (60–100)	88 (55–100)	93 (55–100)

a   *Glyceria maxima* sub-community
b   *Alisma plantago-aquatica-Sparganium erectum* sub-community
5   *Glycerietum maximae* (total)

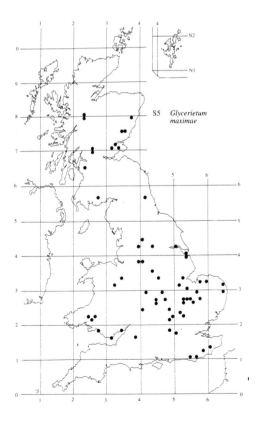

S5   *Glycerietum maximae*

S6
Carex riparia swamp
Caricetum ripariae   Soó 1928

Synonymy
Caricetum acutiformo-ripariae Soó (1927) 1930 *p.p.*;
Caricetum acutiformo-paniculatae Vl. & Van Zinderen
Bakker 1942 *p.p.*; *Angelico-Phragmitetum caricetosum
ripariae* Ratcliffe & Hattey 1982.

Constant species
Carex riparia.

Physiognomy
The *Caricetum ripariae* is generally dominated by *Carex
riparia*, the large tufts of which carry erect leaves forming a sometimes dense canopy usually more than 1 m
tall. The vegetation is typically rather species-poor and
pure stands are not uncommon. Frequently, however,
the community is marked by the patchy abundance of
other swamp emergents and/or tall herbs. Among these,
*Phragmites australis, Equisetum fluviatile, E. palustre,
Epilobium hirsutum, Phalaris arundinacea* and *Filipendula ulmaria* are the most frequent and there is occasionally prominent *Galium palustre* and *Mentha aquatica.*
Less frequent, though sometimes conspicuous, are *Sparganium erectum, Typha latifolia, Juncus effusus, Lycopus
europaeus, Oenanthe crocata, Solanum dulcamara* and
Carex acuta.

Habitat
The community seems to be most characteristic of wet or
waterlogged, mesotrophic to eutrophic, circumneutral
mineral soils alongside standing or slow-moving waters.
It occurs, sometimes as large stands, by sluggish rivers
and streams, in drainage ditches, around ponds and
lakes and in clearings within fen woodlands, always in
the lowlands.
   The *Caricetum ripariae* has been observed growing as
emergent vegetation in up to 20 cm of water but, even
where the water-table falls below the surface, the substrate very typically has an upper layer of very sloppy
sapropelic silt. pH values of 5.8–7.0 have been recorded.

Zonation and succession
The community often forms part of riparian sequences
which frequently terminate abruptly above in agricultural boundaries. Stands within wet woodlands may
pass gradually to the surrounding vegetation with the *C.
riparia* remaining a prominent component of the field
layer.

Distribution
The *Caricetum ripariae* is typically a community of the
agricultural lowlands of England and Wales but,
although *C. riparia* retains an extensive distribution
over much of this ground (Jermy *et al.* 1982), extensive
stands seem to be declining in the Midlands and East
Anglia (C. D. Pigott, personal communication). In
Wales, the community remains quite widespread in the
southern coastal lowlands and is especially distinctive of
the valley and flood-plain mires of Pembrokeshire (Ratcliffe & Hattey 1982).

Affinities
Swamp dominated by *C. riparia* has rarely been referred
to in the British literature but the community as defined
here corresponds to the *Caricetum ripariae* described
from elsewhere in Europe (Soó 1928, Westhoff & den
Held 1969). *C. riparia* is also a locally abundant consti-
tuent of certain kinds of fen vegetation, for example, the
Glyceria sub-community of the *Peucedano-Phragmitetum* and the *Phragmites-Urtica* community, in both of
which it occurs with some tall-fen herbs, as here, but
with other bulky monocotyledons more prominent.

Floristic table S6

Carex riparia	V (6–10)
Epilobium hirsutum	II (1–7)
Phalaris arundinacea	II (1–7)
Equisetum palustre	II (1–3)
Filipendula ulmaria	II (1–7)
Phragmites australis	II (1–6)
Galium palustre	II (1–7)
Equisetum fluviatile	II (1–7)
Mentha aquatica	II (1–5)
Lycopus europaeus	I (3–7)
Sparganium erectum	I (6–7)
Juncus effusus	I (1–5)
Urtica dioica	I (1–7)
Oenanthe crocata	I (1–7)
Typha latifolia	I (1–7)
Solanum dulcamara	I (1–7)
Carex acuta	I (1–7)
Juncus acutiflorus	I (1–5)
Galium aparine	I (1–7)
Agrostis stolonifera	I (1–7)
Callitriche stagnalis	I (1–5)
Calystegia sepium	I (6)
Eupatorium cannabinum	I (1–4)
Polygonum amphibium	I (1–3)
Angelica sylvestris	I (1–3)
Calliergon giganteum	I (1–3)
Poa trivialis	I (3)
Scutellaria galericulata	I (1–3)
Rubus fruticosus agg.	I (2)
Myosotis scorpioides	I (5)
Rumex hydrolapathum	I (4)
Nymphaea alba	I (5)
Typha angustifolia	I (4)
Cirsium palustre	I (1–3)
Rumex crispus	I (2)
Nasturtium officinale	I (2)
Carex elata	I (3)
Valeriana officinalis	I (3)
Cardamine pratensis	I (1)
Epilobium palustre	I (1)
Dryopteris dilatata	I (1)
Cirsium arvense	I (1)
Polygonum hydropiper	I (1)
Number of samples	113
Number of species/sample	5 (1–17)
Vegetation height (cm)	128 (100–200)
Vegetation cover (%)	90 (40–100)

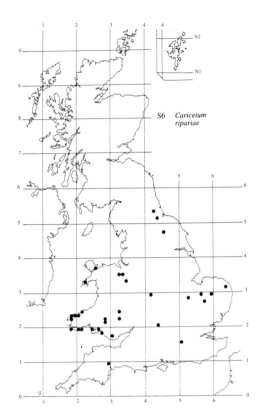

S6   *Caricetum*
      ripariae

Carex acuta in swamps and fens

Carex acuta is a common sedge with a widespread distribution in the lowlands of England, Wales and southern Scotland (Jermy *et al.* 1982) and it is most characteristic of the margins of sluggish or standing, mesotrophic to eutrophic waters. It has, however, only rarely been recorded as the overwhelming dominant in swamp vegetation and a distinct *Caricetum acutae* has not been recognised in our scheme (cf. Tüxen 1937, Westhoff & den Held 1969, Oberdorfer 1977, Birse 1980). Swamp stands in which *C. acuta* is locally abundant have here been classified in the *Phragmitetum australis*, the *Glycerietum maximae* or the *Caricetum ripariae*. The species may also be prominent in some stands of the *Phragmites-Eupatorium* and *Phragmites-Urtica* fens.

C. acuta can be confused in its vegetative state with *C. acutiformis* and the two may hybridise. Further sampling is needed to ascertain its distribution in pure swamp stands.

S7
Carex acutiformis swamp
Caricetum acutiformis   Sauer 1937

Synonymy
Caricetum acutiformo-ripariae Soó (1927) 1930 *p.p.*; *Caricetum acutiformo-paniculatae* Vl. & Van Zinderen Bakker 1942 *p.p.*; *Carex acutiformis* fen Lambert 1951 *p.p.*; *Carex acutiformis* stands Meres Survey 1980 *p.p.*

Constant species
Carex acutiformis.

Physiognomy
The *Caricetum acutiformis* is always dominated by *Carex acutiformis* forming an open or closed canopy of shoots and arcuate leaves about 1 m tall. No other species is constant but there are usually some scattered tall fen herbs such as *Angelica sylvestris* and *Valeriana officinalis* and shorter species like *Galium palustre* and *Mentha aquatica*. Other swamp species, e.g. *Carex paniculata*, *Sparganium erectum* and *Typha latifolia*, may be locally prominent, and *Juncus effusus* is sometimes abundant. However, many of the occasionals reflect the particular floristic context of the often small stands.

Habitat
The community seems to be typical of situations which are, in some respects, similar to those occupied by the *Caricetum ripariae*. It has been recorded from moderately eutrophic, circumneutral substrates on the margins of slow-moving or standing lowland waters in open-water transitions, in wet hollows within flood-meadows, in ditches and alongside sluggish streams and rivers. Here the water-level may be up to about 20 cm above ground and the substrate pH 6.0–6.8. There is some evidence, however, (e.g. Haslam 1978), that the *Caricetum acutiformis* is more consistently associated with calcareous habitats than is the *Caricetum ripariae*: it occurs, for example, in ditches in fen peat and also on the margins of slow chalk streams.

Zonation and succession
Around more extensive open-water transitions, the community may form swamp which passes gradually to fen in which *C. acutiformis* remains a prominent component, e.g. some forms of the *Peucedano-Phragmitetum*, and it was from such situations that Lambert (1951) described her *C. acutiformis* sere along the Bure valley in Norfolk. At more abrupt water margins, the community occurs in often narrow and fragmentary transitions with the *Sparganietum erecti* towards deeper and, to landward, the *Glycerietum maximae* or *Phragmitetum australis*. Unlike the *Caricetum ripariae*, this community may also form swampy patches in calcareous flood-meadows and flood-pastures, passing gradually through some form of Calthion community to damp mesotrophic grassland, but, with agricultural improvement, such transitions are becoming more rare. *C. acutiformis* seems to be quite an aggressive species which may be able to readily invade riverside fields where drains become blocked. It also appears to be able to tolerate cattle grazing (Wheeler 1975).

Distribution
C. acutiformis is not so obviously restricted to the south and east as is *C. riparia* (Jermy *et al.* 1982), although, like that species, it is primarily a lowland sedge. The *Caricetum acutiformis* swamp is, however, not a common community and it has been encountered at scattered localities, notably in the Fens and Broads and around the Shropshire meres.

Affinities
Like *C. riparia*, *C. acutiformis* may also be a prominent component of both fen and fen woodland vegetation but the stands included here are distinct in their species-poverty and overwhelming dominance of the sedge. The *Caricetum acutiformis* also shows affinities with certain Calthion communities where, with other sedges, rushes and poor-fen dicotyledons, *C. acutiformis* forms a species-rich sward on gleyed soils.

Floristic table S7

Carex acutiformis	V (7–10)
Juncus effusus	III (2–4)
Galium palustre	II (1–3)
Mentha aquatica	II (3–7)
Lotus uliginosus	II (2–3)
Arrhenatherum elatius	II (1–2)
Valeriana officinalis	II (1–3)
Angelica sylvestris	II (3)
Solanum dulcamara	I (2)
Cardamine amara	I (4)
Holcus lanatus	I (3)
Rumex crispus	I (4)
Equisetum palustre	I (4)
Filipendula ulmaria	I (3)
Poa trivialis	I (4)
Carex paniculata	I (4)
Caltha palustris	I (3)
Lemna minor	I (3)
Cicuta virosa	I (2)
Ranunculus acris	I (2)
Scutellaria galericulata	I (2)
Ranunculus repens	I (2)
Epilobium hirsutum	I (2)
Juncus inflexus	I (2)
Lythrum salicaria	I (3)
Polygonum aviculare	I (5)
Phalaris arundinacea	I (2)
Sparganium erectum	I (2)
Stellaria alsine	I (3)
Symphytum officinale	I (4)
Typha latifolia	I (2)
Cirsium palustre	I (1)
Anthoxanthum odoratum	I (1)
Galium aparine	I (1)
Festuca rubra	I (1)
Lathyrus pratensis	I (1)
Polygonum hydropiper	I (1)
Urtica dioica	I (1)
Veronica beccabunga	I (1)
Brachythecium rivulare	I (1)
Number of samples	5
Number of species/sample	11 (6–25)
Vegetation height (cm)	98 (70–120)
Vegetation cover (%)	98 (90–100)

S8
Scirpus lacustris ssp. lacustris swamp
Scirpetum lacustris (Allorge 1922) Chouard 1924

Synonymy

Scirpus lacustris open reed-swamp auct. angl.; Scirpo-Phragmitetum Koch 1926 p.p.. (not sensu Wheeler 1980a); Scirpeto-Phragmitetum medioeuropaeum (Koch 1926) R.Tx. 1941 p.p.

Constant species

Scirpus lacustris ssp. lacustris.

Physiognomy

The Scirpetum lacustris typically has a somewhat open cover of S. lacustris ssp. lacustris, the stout emergent flowering shoots of which may reach more than 2 m above water-level. The robust creeping rhizomes frequently produce tufts of totally submerged leaves. No other species is frequent throughout but the sub-communities are distinguished by preferentially frequent species which may attain local abundance. It should be noted that, in the data available, submerged and floating-leaved aquatics are largely confined to the Equisetum fluviatile sub-community. This is probably, at least in part, a reflection of the fact that in many lowland waters, the sometimes rich aquatic flora which often overlapped in distribution with the Scirpetum lacustris has been lost because of pollution and disturbance.

Sub-communities

Scirpus lacustris ssp. lacustris sub-community: Scirpus lacustris open reed-swamp Pallis 1911; Scirpetum lacustris, Scirpus-Nymphaea mictium and Phragmites-Scirpus associes p.p.. Tansley 1939; Schoenoplectus lacustris society Wheeler 1978; Association of Phragmites australis and Schoenoplectus lacustris Pigott & Wilson 1978. Here are included pure stands of S. lacustris ssp. lacustris and mixed stands with P. australis where the former species remains the more abundant. No other species is frequent or abundant, although it is clear that, in the past, such swamps were sometimes associated with a rich aquatic flora including Nuphar lutea, Nymphaea alba, Lemna minor, L. gibba, Wolfia arrhiza, Spirodela polyrhiza, Potamogeton natans, P. lucens, P. perfoliatus and P. crispus (e.g. Pallis 1911, Tansley 1939).

Sparganium erectum sub-community: Scirpus lacustris riparian reedswamp Butcher 1933. In this sub-community, there is a generally lower emergent canopy of S. lacustris ssp. lacustris intermixed with Sparganium erectum and, sometimes, Alisma plantago-aquatica.

Equisetum fluviatile sub-community: Schoenoplectus lacustris-Juncus fluitans sociation Spence 1964; Schoenoplectus lacustris-Phragmites communis Association, Schoenoplectus lacustris-Equisetum fluviatile subassociation Birks 1973. Here, there is again a generally shorter canopy of S. lacustris ssp. lacustris but, in this sub-community, the shoots are intermixed with emergent Equisetum fluviatile, Carex rostrata and Menyanthes trifoliata, each of which may be locally prominent. Less frequent are Potentilla palustris, Galium palustre and Ranunculus flammula and, with floating leaves, Potamogeton natans and Nymphaea alba. Juncus bulbosus (including J. fluitans Lam.) occasionally forms submerged tangles and, in some Scottish lochs, Sparganium minimum has been reported as a distinctive associate (Spence 1964). A wide variety of other aquatic and fen species occur rarely.

Habitat

The Scirpetum lacustris is, above all, a swamp of deep water occurring typically in larger pools and lakes, in high-order streams and, more rarely, in canals and dykes. It is predominantly a lowland vegetation type.

The community has been recorded from water depths up to 150 cm and it does not seem to occur where there is less than 25 cm of water. Its smooth and elastic stems may offer less resistance than other swamp emergents to the wave turbulence and high winds that can develop over extensive stretches of deeper, open water (Tansley

1939). It is most characteristic of standing or slow-moving waters although it can resist fast flows, provided there is not much spate which tends to snap off the emergent shoots (Haslam 1978). In faster rivers, the proportion of submerged leaves increases, though these too can be damaged by much turbulence (Butcher 1933, Haslam 1978).

S. lacustris ssp. *lacustris* has both superficial creeping rhizomes which form a robust network and a mass of more deeply penetrating roots and it has been suggested (Haslam 1978) that this might account for its ability to colonise a variety of substrates from fine but root-resistant stable gravels to soft deep silts. Pearsall (1921) considered that, in the Lake District, the *Scirpetum* was typical of more organic substrates than those occupied by the *P. australis* swamps.

The floristic differences between the sub-communities seem to be related to water depth and the trophic state of the habitat. The *Sparganium* sub-community is typically encountered in shallower, more eutrophic waters in slow-flowing clay streams and silted dykes where there may be additional enrichment by run-off from surrounding farmland. The *Equisetum* sub-community, on the other hand, is characteristic of deeper and more nutrient- and base-poor waters typical of lakes over hard arenaceous rocks towards the upland fringes. The *Scirpus* sub-community occurs in a variety of situations where the waters are of intermediate quality, although it also includes very species-poor stands in deeper waters in both oligotrophic and eutrophic sites.

Zonation and succession

The *Scirpetum lacustris* represents the deep-water limit of swamp vegetation in Britain and stands may be isolated beyond the more proximal parts of emergent sequences. In larger lakes, where the community is represented by the *Scirpus* or, in shallower water, the *Equisetum* sub-community, extensive stretches of open water may occur between the community and inshore swamps. In Scottish lakes, Spence (1964) commonly encountered mixtures of submerged *Juncus bulbosus* (i.e. *J. fluitans* Lam.) and *Littorella uniflora* in this intervening zone and often running out to form, with *Sparganium minimum*, a distinctive understorey to the *Scirpetum*. In other cases, the community gave way, in shallower water, to *Phragmitetum australis*, sometimes with a narrow zone of overlap between the two vege-

tation types. An essentially similar zonation occurs at Esthwaite in Cumbria (Pearsall 1918, Tansley 1939, Pigott & Wilson 1978).

It is clear that, at North Fen at Esthwaite, the *Scirpetum* advanced some 15–30 m into open water between 1915 and 1929 (Pearsall 1918, Tansley 1939) and, along part of its front, about half this distance between 1929 and 1967 (Pigott & Wilson 1978), giving rates of movement between about 0.5 and 2 m yr^{-1}. At other sites, however, stands have remained more or less unaltered and stationary for at least 50 years (West 1905, Spence 1964: cf. Plates 70 & 71 in Spence 1964) and *S. lacustris* ssp. *lacustris* may itself be of little importance in terrestrialisation.

In rivers and streams, small and fragmentary stands of the *Scirpus* sub-community may stand some distance out from the banks in deeper water. On more steeply shelving profiles, the *Sparganium* sub-community may form part of a narrow transition from open water, through the *Sparganietum erecti* to the *Glycerietum maximae* or some form of tall-herb vegetation (Haslam 1978).

Distribution

S. lacustris ssp. *lacustris* is predominantly a lowland species, being commonest in the Midlands and southern England but, in this region, the *Scirpetum* generally occurs as small stands of the *Scirpus* or *Sparganium* sub-communities. Larger stands of the *Scirpus* or *Equisetum* sub-communities are more characteristic of the northern and west lowlands where *S. lacustris* ssp. *lacustris* is more locally distributed but where larger bodies of open water are more frequent. The *Scirpetum* is notably uncommon in Broadland (Lambert 1946, 1951; Wheeler 1978).

Affinities

Unlike many other emergents, *S. lacustris* ssp. *lacustris* occurs but sparsely in communities other than the swamp vegetation in which it is the dominant and most of the difficulties of classification arise because of its partial overlap in shallower waters with *P. australis*. In some schemes, vegetation with either or both of these species dominant has been placed in a single large swamp type such as the *Scirpeto-Phragitetum* (Koch 1926, Tüxen 1941).

Floristic table S8

	a	b	c	8
Scirpus lacustris lacustris	V (5–10)	V (5–10)	V (8–9)	V (5–10)
Phragmites australis	II (1–3)			I (1–3)
Lythrum salicaria	I (3)			I (3)
Sparganium erectum		V (3–8)		II (3–8)
Alisma plantago-aquatica		II (4–6)		I (4–6)
Lemna minor		I (3)		I (3)
Hydrocharis morsus-ranae		I (2)		I (2)
Ceratophyllum demersum		I (6)		I (6)
Elodea canadensis		I (6)		I (6)
Lemna gibba		I (9)		I (9)
Sagittaria sagittifolia		I (6)		I (6)
Juncus conglomeratus		I (4)		I (4)
Lycopus europaeus		I (3)		I (3)
Spirodela polyrhiza		I (9)		I (9)
Polygonum bistorta		I (3)		I (3)
Solanum nigrum		I (4)		I (4)
Typha latifolia		I (5)		I (5)
Equisetum fluviatile			V (2–6)	II (2–6)
Carex rostrata			IV (3–4)	II (3–4)
Menyanthes trifoliata			III (3–4)	II (3–4)
Potamogeton natans			III (2–4)	II (2–4)
Nymphaea alba		I (1)	II (1–3)	I (1–3)
Juncus bulbosus			II (2–4)	I (2–4)
Ranunculus flammula			II (1–2)	I (1–2)
Potentilla palustris			II (3)	I (3)
Galium palustre			II (2–3)	I (2–3)
Juncus acutiflorus			I (2)	I (2)
Utricularia minor			I (1)	I (1)
Potamogeton lucens			I (1)	I (1)
Potamogeton polygonifolius			I (4)	I (4)
Myriophyllum alterniflorum			I (2)	I (2)
Potamogeton perfoliatus			I (1)	I (1)
Sparganium angustifolium			I (1)	I (1)
Carex curta			I (4)	I (4)
Caltha palustris			I (1)	I (1)
Angelica sylvestris			I (1)	I (1)
Epilobium palustre			I (2)	I (2)
Hydrocotyle vulgaris			I (2)	I (2)
Agrostis canina canina			I (2)	II (2)
Carex nigra			I (1)	I (1)
Filipendula ulmaria			I (4)	I (4)
Juncus effusus			I (2)	I (2)
Mentha aquatica			I (4)	I (4)
Polygonum amphibium			I (2)	I (2)
Stachys palustris			I (4)	I (4)

	a	b	c	8
Solanum dulcamara	I (2)	I (3)		I (2–3)
Glyceria fluitans	I (2)		I (1)	I (1–2)
Eleocharis palustris	I (3)		I (4)	I (3–4)
Iris pseudacorus		I (1)	I (5)	I (1–5)
Nuphar lutea		I (4)	I (5)	I (4–5)
Number of samples	5	5	8	18
Number of species/sample	2 (1–3)	6 (2–8)	8 (4–13)	5 (1–13)
Vegetation height (cm)	180 (150–210)	130 (60–230)	112 (100–124)	135 (60–230)
Vegetation cover (%)	60 (20–100)	99 (95–100)	84 (40–100)	81 (20–100)

a *Scirpus lacustris lacustris* sub-community
b *Sparganium erectum* sub-community
c *Equisetum fluviatile* sub-community
8 *Scirpetum lacustris* (total)

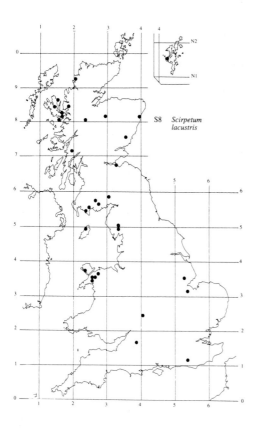

S8 *Scirpetum lacustris*

S9

Carex rostrata swamp
Caricetum rostratae   Rübel 1912

Synonmy
Carex ampullacea consocies   Matthews 1914; *Carice-tum inflatae* Tansley 1939; *Carex rostrata* 'reed-swamps' Holdgate 1955*b* *p.p.*; *Carex rostrata* socia-tions Spence 1964 *p.p.*; *Carex rostrata* reed-swamps Proctor 1974; *Carex rostrata* nodum Daniels 1978; *Caricetum rostratae* Birse 1980 *p.p.*

Constant species
Carex rostrata.

Rare species
Eriocaulon septangulare.

Physiognomy
The *Caricetum rostratae* is generally dominated by *Carex rostrata* which characteristically forms a some-what open cover of tufted shoots usually 50–60 cm tall. No other species is frequent throughout and the vege-tation is typically species-poor.

Sub-communities

***Carex rostrata* sub-community.** Open *Carex rostrata* sociation Spence 1964. This sub-community includes pure and very species-poor stands overwhelmingly dominated by *C. rostrata*. *Equisetum fluviatile*, *Poly-gonum amphibium* and *Potamogeton natans* occur occasionally.

***Menyanthes trifoliata-Equisetum fluviatile* sub-commun-ity:** *Carex rostrata-Menyanthes* sociation Spence 1964; *Carex rostrata-Menyanthes trifoliata* Association Birks 1973. Here, the vegetation comprises mixtures of *C. rostrata*, *Equisetum fluviatile*, *Menyanthes trifoliata* and *Potentilla palustris* sometimes developed as a float-ing mat. Although the sedge is generally dominant, each of these associates may be locally abundant, the bulky foliage of *M. trifoliata* and *P. palustris* often appearing particularly prominent among the thinner sedge and

horsetail shoots. *Eleocharis palustris*, *Carex nigra*, *Ranunculus flammula*, *Caltha palustris* and *Potamogeton polygonifolius* are occasional. *Lobelia dortmanna* and *Littorella uniflora* are uncommon, though sometimes abundant and, on Skye, *Eriocaulon septangulare* occurs in this vegetation.

Habitat
The *Caricetum rostratae* is typically a swamp of shallow to moderately deep, mesotrophic to oligotrophic, stand-ing waters with organic substrates. Although found down almost to sea-level, it is one of the few swamp communities that makes a major contribution to the vegetation of upland lakes where stands may be exten-sive. It also occurs more fragmentarily in peat cuttings.

Although the community can be encountered on silty or sandy substrates, it is more typical of an organic base, often being rooted directly in firm peat (as where the *Carex* sub-community is colonising existing underwater deposits) or spongey peat ooze (especially under the *Menyanthes-Equisetum* sub-community which produces abundant litter). pH values of 5.0–6.8 have been recorded but the waters may be nutrient-poor and the *Caricetum rostratae* includes stands which extend the occurrence of swamp vegetation into highly oligotro-phic situations.

The two sub-communities are associated with differ-ent ranges of water depth, the *Carex* sub-community occurring in as much as 1 m of water and rarely in less than 10 cm (mean about 30 cm), the *Menyanthes-Equisetum* sub-community in 2–40 cm (mean about 20 cm).

Zonation and succession
Sometimes, especially in the more oligotrophic upland lakes, the community represents the limit of swamp vegetation. Frequently, however, it occurs behind a front of the *Scirpetum lacustris* and/or *Phragmitetum australis*. The two sub-communities are themselves com-monly zoned, with the *Carex* sub-community extending

out into deeper water and giving way behind to the *Menyanthes-Equisetum* sub-community. This may grade laterally to the *C. rostrata* sub-community of the *Equisetetum fluviatile* with a switch in dominance to *E. fluviatile* and, in shallow water around some Scottish lakes, to the *Caricetum vesicariae* in which *C. vesicaria* is dominant with many of the same associates.

In some cases, this kind of transition continues above in a gradual switch to the *Potentillo-Caricetum* with an increase in poor-fen herbs and larger *Calliergon* spp. (see also Matthews 1914) and there seems little doubt that this represents a standard kind of succession in fairly base-poor waters over organic substrates. Sometimes, however, this kind of smooth zonation is complicated by local variation in water throughput and enrichment around inflows and along seepage lines. Then, the *Caricetum* may be part of a complex patchwork of poor fens in which *C. rostrata* remains prominent but where the understorey varies in response to an increase in calcium and base-status, as in the *C. rostrata-Sphagnum squarrosum* community and certain types of *Carex rostrata-Calliergon* mire vegetation. This kind of rich local variation has been well described from some basin mires on Carboniferous Limestone, such as Malham Tarn in North Yorkshire (Proctor 1974, Adam *et al.* 1975) and Sunbiggin Tarn, Cumbria (Holdgate 1955*b*). It is also seen at the margins of some of the Scottish lakes (Spence 1964). In such situations, other sedges such as *C. lasiocarpa* and *C. aquatilis* may attain local prominence in standing water alongside *C. rostrata*.

Where the *Caricetum rostratae* occurs in pools within non-calcareous basin mires or in peat cuttings in ombrogenous mires, it may grade to the more oligotrophic vegetation of the *C. rostrata-Sphagnum recurvum* community of the stagnant bog-pool margins.

Distribution
The *Caricetum rostratae* is very much a community of the north and west with a very few outliers in the southern and eastern lowlands.

Affinities
There are two difficulties in characterising *C. rostrata* vegetation of this kind. The first is to decide whether and how to separate a *Caricetum rostratae sensu stricto* from other vegetation in which *C. rostrata* and, to a lesser extent, *Menyanthes trifoliata* and *Potentilla palustris* remain frequent and abundant. This question is complicated by the fact that the development of a floating mat in various of these vegetation types means that there is sometimes no hard and fast physiognomic distinction between swamp and fen. Here the absence of any prominent bryophyte layer, whether of larger *Calliergon* spp., Sphagna or 'brown mosses', and the general infrequency of poor-fen herbs such as *Cardamine pratensis*, *Galium palustre* and *Epilobium palustre*, are taken as distinguishing features of a *Caricetum rostratae*. Other workers (e.g. Birse 1980) have included in a *Caricetum rostratae* vegetation which is here placed in the *Potentillo-Caricetum*.

Second, there is the problem of marking off a *Caricetum rostratae* from vegetation in which *Equisetum fluviatile*, *M. trifoliata*, *P. palustris* and, to a lesser extent, *C. rostrata* remain frequent with a variety of other dominants such as *Carex vesicaria*, *C. aquatilis*, *C. lasiocarpa*, *Scirpus lacustris* ssp. *lacustris*, *Typha latifolia* and *P. australis*. Here, the dominance of *C. rostrata* is taken as a distinguishing feature of the *Caricetum rostratae*, although there are situations where it is difficult to allocate stands on this basis. Separations between this community and the *Equisetetum fluviatile* are particularly problematic.

Floristic table S9

	a	b	9
Carex rostrata	V (5–10)	V (6–9)	V (5–10)
Polygonum amphibium	II (1–4)		I (1–4)
Potamogeton natans	II (2–7)	I (3–6)	I (3–7)
Mentha aquatica	I (1–2)		I (1–2)
Juncus effusus	I (2–3)		I (2–3)
Equisetum fluviatile	II (1–4)	IV (1–6)	III (1–6)
Menyanthes trifoliata	I (1–5)	IV (1–8)	II (1–8)
Potentilla palustris	I (1)	III (1–6)	II (1–6)
Eleocharis palustris	I (1–4)	II (3–4)	I (1–4)
Potamogeton polygonifolius		II (1–8)	I (1–8)
Caltha palustris		II (1–4)	I (1–4)
Ranunculus flammula		II (1–4)	I (1–4)
Carex nigra		II (1–5)	I (1–5)
Cardamine pratensis		I (1–2)	I (1–2)
Agrostis stolonifera		I (1–3)	I (1–3)
Epilobium palustre		I (1–3)	I (1–3)
Calliergon giganteum		I (1–3)	I (1–3)
Juncus acutiflorus		I (1–7)	I (1–7)
Lobelia dortmanna		I (3–5)	I (3–5)
Pedicularis palustris		I (1–4)	I (1–4)
Littorella uniflora		I (4–5)	I (4–5)
Eriocaulon septangulare		I (2–5)	I (2–5)
Juncus bulbosus		I (1–3)	I (1–3)
Utricularia vulgaris		I (1)	I (1)
Hydrocotyle vulgaris	I (1–2)	I (1–2)	I (1–2)
Juncus articulatus	I (1)	I (1–2)	I (1–2)
Nymphaea alba	I (7)	I (2–4)	I (2–7)
Myosotis laxa caespitosa	I (3)	I (1)	I (1–3)
Galium palustre	I (3)	I (1)	I (1–3)
Number of samples	21	31	52
Number of species/sample	4 (1–10)	7 (3–17)	6 (1–17)
Vegetation height (cm)	63 (25–83)	57 (25–90)	60 (25–90)
Vegetation cover (%)	70 (35–100)	69 (30–100)	69 (30–100)

a   *Carex rostrata* sub-community
b   *Menyanthes trifoliata-Equisetum fluviatile* sub-community
9   *Caricetum rostratae* (total)

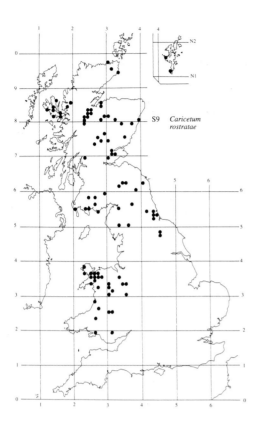

S9 *Caricetum rostratae*

Carex lasiocarpa in swamps and fens

Carex lasiocarpa has a widespread distribution to the north and west of Britain with some records for lowland England, mainly in East Anglia (Jermy *et al.* 1982). It is most characteristic of wet, mesotrophic conditions but, although it has sometimes been described as forming pure stands in open-water transitions (e.g. Spence 1964, Jermy *et al.* 1982), very few samples are available to form the basis of a *Caricetum lasiocarpae* swamp. Published descriptions of vegetation dominated by this sedge (e.g. Spence 1964, Birse 1980) are based on rather heterogeneous data which is here considered best allocated to a number of other communities.

Swamp vegetation in which *C. lasiocarpa* is locally prominent is accommodated in this scheme in the *Phragmitetum australis* and the *Menyanthes* sub-community of the *Cladietum marisci*. The species is also an important component of some stands of the *Cicuta* sub-community of the *Peucedano-Phragmitetum* rich fen. In general, however, the locus of this species in Britain, as on the Continent, is not within the swamps and fens of the Phragmitetea but in the small-sedge mires of the Parvocaricetea, particularly in the brown-moss mires of the Caricion davallianae. *C. lasiocarpa* is abundant in some stands of the *Carex rostrata-Calliergon* fen and occurs occasionally in a number of other communities in that alliance.

Carex aquatilis in swamps and fens

Carex aquatilis has an Arctic–Subarctic distribution, being limited in Britain to Scotland and a few outlying stations in the Lake District and Wales (Matthews 1955, Jermy *et al.* 1982). It is characteristic of two distinct kinds of vegetation. The first is swamp or very wet fen in open-water transitions around more mesotrophic lakes to the north and west. Here it often occurs with, or is a local replacement for, *Carex vesicaria* and *C. rostrata* and stands in which it is locally prominent in this way have here been allocated to the *Caricetum vesicariae*

swamp and the *Potentillo-Caricetum rostratae* fen. A separate *Caricetum aquatilis* has not been distinguished and the *Lysimachio-Caricetum aquatilis* described from Scotland by Birse (1980) can be accommodated comfortably within the *Potentillo-Caricetum*.

The very different montane *Carex curta-Sphagnum russowii* small-sedge mire in which *C. aquatilis* occurs with *C. rariflora*, *C. curta* and a variety of small herbs in a *Sphagnum* carpet is described among the mires of Volume 2 of *British Plant Communities*.

S10
Equisetum fluviatile swamp
Equisetetum fluviatile Steffen 1931 *emend.* Wilczek 1935

Synonymy
Equisetum limosum reedswamp Rankin 1911; *Equisetum fluviatile* reedswamp Tansley 1939; *Scirpeto-Phragmitetum medioeuropaeum* (Koch 1926) R.Tx. & Preising 1942 *p.p.*

Constant species
Equisetum fluviatile.

Rare species
Calamagrostis stricta, Lysimachia thyrsiflora.

Physiognomy
The *Equisetetum fluviatile* comprises open or closed vegetation up to about 50 cm high in which *Equisetum fluviatile* is generally the most abundant species. No other species is frequent throughout, although in each of the sub-communities some of the associates may be locally abundant and their prominence is often emphasised by the thin shoots of the 'dominant'.

Sub-communities

Equisetum fluviatile sub-community: Open *Equisetum fluviatile* sociation Spence 1964; Sociatie van *Equisetum fluviatile* Westhoff & den Held 1969. Here are included pure and very species-poor stands in which *E. fluviatile* is overwhelmingly the most abundant species. Occasionals include species of periodically inundated finer sediments such as *Polygonum hydropiper* and *Rorippa islandica* and, around Scottish lakes, *Littorella uniflora* has been reported as a common associate of this kind of vegetation (Spence 1964).

Carex rostrata sub-community. In the richer vegetation of this sub-community the shoots of *E. fluviatile* occur intermixed with tufts of *C. rostrata*, although the former is always the more abundant. *Menyanthes trifoliata* and *Potentilla palustris* are constant as an understorey and,

on occasion, may dominate. Within this mat, which sometimes occurs as a swinging semi-submerged vegetation, there may be scattered plants of *Galium palustre*, *Epilobium palustre* and *Eriophorum angustifolium*. *Calamagrostis stricta* and *Lysimachia thyrsiflora* have been recorded here.

Habitat
Both sub-communities can occur in similar situations to the *Caricetum rostratae*, being found in shallow to moderately deep, eutrophic to oligotrophic, standing waters in both lowland and upland lakes and pools. Here, the water can be up to more than 1 m deep with a sediment pH of 5.2–6.4. The *Equisetetum*, however, seems to be as characteristic of silty and sandy substrates as of peaty deposits and the *Equisetum* sub-community occurs in habitats where the *Caricetum rostratae* is very rarely found: on fine inorganic material around the draw-down zone of reservoirs and the inundated margins of lowland pools and very slack reaches of high-order streams.

Zonation and succession
In open-water transitions of larger lakes, especially where nutrient-poor waters occur over organic substrates, the community occurs in similar zonations to those involving the *Caricetum rostratae* and it commonly grades laterally to that community with a switch in dominance to *C. rostrata*. On more inorganic material in such situations, it may also occur alongside the *E. fluviatile* sub-community of the *Eleocharitetum palustris* (e.g. Spence 1964).

Around the margins of reservoirs and lowland pools with inorganic substrates, the *Equisetum* sub-community often forms a zone, sometimes with the *Eleocharis* sub-community of the *Eleocharitetum*, between open water and Elymo-Rumicion inundation communities or, where stock water, poached Cynosurion swards (Figure 13).

Distribution

The *Equisetetum* occurs over much the same range as the *Caricetum rostratae*, primarily in the north and west where large fairly oligotrophic water bodies are common, but the *Equisetum* sub-community extends the distribution into the eastern and southern lowlands where small stands are widespread.

Affinities

A distinct *Equisetetum* has rarely been separated off from *Carex rostrata* swamp vegetation and indeed there is a complete gradation between the two communities as

defined here, in both the presence and absence of the distinctive asssociates *Menyanthes trifoliata* and *Potentilla palustris*. To a lesser extent, the *Equisetetum* also grades floristically to the *Eleocharitetum palustris* and elements of the community may form an understorey to swamps with larger dominants such as *Phragmites australis* and *Scirpus lacustris* ssp. *lacustris*. Separations between these vegetation types based on abundance are not helped by the slim nature of the aerial parts of *E. fluviatile* which, even when very abundant, do not create the impression of physiognomic dominance.

Figure 13. Mosaic of aquatic and inundation communities, swamps, fen and grassland over the draw-down zone and inlet streams of a reservoir in County Durham.

A10   *Polygonum amphibium* community in aquatic and amphibious forms
S10   *Equisetetum fluviatile* swamp
S19   *Eleocharitetum palustris* swamp
S28   *Phalaridetum arundinaceae* tall-herb fen
MG9   *Holcus-Deschampsia* grassland
WE34   *Polygonum persicaria-Polygonum lapathifolium* inundation community with dense stands of *Juncus filiformis* around the margins

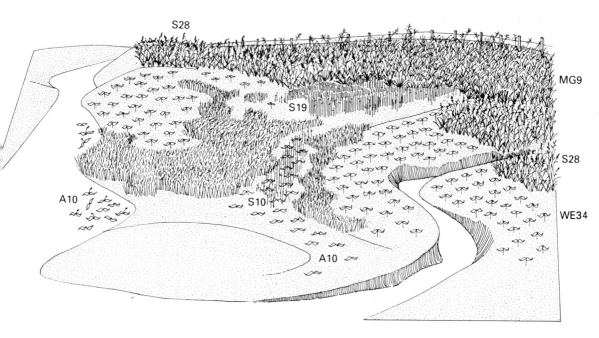

Floristic table S10

	a	b	10
Equisetum fluviatile	V (8–10)	V (6–8)	V (6–10)
Polygonum hydropiper	II (2–4)		I (2–4)
Solanum dulcamara	I (4–5)		I (4–5)
Rorippa islandica	I (2–4)		I (2–4)
Polygonum amphibium	I (3)		I (3)
Alisma plantago-aquatica	I (1)		I (1)
Callitriche stagnalis	I (3–8)		I (3–8)
Lemna minor	I (3)		I (3)
Ranunculus flammula	I (3–4)		I (3–4)
Carex rostrata		V (2–5)	II (2–5)
Menyanthes trifoliata		IV (5–9)	II (5–9)
Potentilla palustris		IV (2–7)	II (2–7)
Galium palustre	I (2)	III (1–4)	II (1–4)
Epilobium palustre	I (1)	II (1–5)	I (1–5)
Eriophorum angustifolium		II (1–5)	I (1–5)
Caltha palustris		I (2–4)	I (2–4)
Angelica sylvestris		I (1)	I (1)
Calliergon cordifolium		I (3)	I (3)
Number of samples	18	7	25
Number of species/sample	4 (1–12)	11 (6–15)	6 (1–15)

a   *Equisetum fluviatile* sub-community
b   *Carex rostrata* sub-community
10 *Equisetetum fluviatile* (total)

there may be a zone of the *Glycerietum maximae* or an abrupt transition to tall-herb vegetation.

On the salt-marshes at Bridgwater and Berrow in Somerset, the *Typha* sub-community occurs in association with the *Scirpetum maritimi*.

Although the association between the *Typhetum* and sites with active silt accretion is commonplace, there is very little systematic information available about its role in succession. At North Fen, Esthwaite, in Cumbria, the community has spread over 70 years from its confines on the rapidly accumulating material around the mouth of the Black Beck to form a belt between the *Phragmitetum australis* swamp and the *Potentillo-Caricetum* (Pearsall 1918, Tansley 1939, Pigott & Wilson 1978).

Distribution

The community is widespread through the agricultural lowlands of England, being less common in Wales and Scotland.

Affinities

T. latifolia is an infrequent component of swamps other than the one in which it is dominant and the affinities of the *Typhetum* are unclear. In Britain, *T. latifolia* and *T. angustifolia* seem rarely to occur together and the combined community *Typhetum angustifolio-latifoliae* (Eggler 1933) Schmale 1939 is inadequate to describe the stands included here.

Floristic table S12

	a	b	c	d	12
Typha latifolia	V (5–10)	V (6–10)	V (7–10)	V (5–9)	V (5–10)
Scirpus lacustris tabernaemontani	I (4)				I (4)
Cicuta virosa	I (1)				I (1)
Aster tripolium	I (1–4)				I (1–4)
Urtica dioica	I (1)				I (1)
Mentha aquatica		V (2–7)	II (2)		II (2–7)
Galium palustre	I (4)	III (2–5)		II (2–4)	II (2–5)
Juncus effusus	I (2–5)	III (1–5)		I (5)	II (1–5)
Myosotis laxa caespitosa		I (3–4)			I (3–4)
Rumex hydrolapathum		I (3–4)			I (3–4)
Berula erecta		I (3–4)			I (3–4)
Juncus acutiflorus		I (3–6)			I (3–6)
Lycopus europaeus		I (3)			I (3)
Rumex crispus		I (3)			I (3)
Acorus calamus		I (3)			I (3)
Scirpus lacustris lacustris		I (4)			I (4)
Alisma plantago-aquatica	I (1–4)	I (3)	IV (1–4)	I (1)	I (1–4)
Sparganium erectum			III (3–4)		I (3–4)
Eleocharis palustris		I (4)	II (4–6)		I (4–6)
Glyceria fluitans			II (1–2)		I (1–2)
Ranunculus sceleratus			II (2–3)		I (2–3)
Callitriche stagnalis			I (4)		I (4)
Lythrum salicaria			I (3)		I (3)
Apium nodiflorum			I (5)		I (5)
Callitriche obtusangula			I (2)		I (2)
Carex rostrata		I (1–5)		V (3–9)	I (1–9)
Menyanthes trifoliata				I (4)	I (4)
Potentilla palustris				I (5)	I (5)
Lemna minor	I (3–10)	II (3–8)	II (5–7)	II (1)	II (1–10)
Hydrocotyle vulgaris	I (3)	I (4–7)		I (3)	I (3–7)

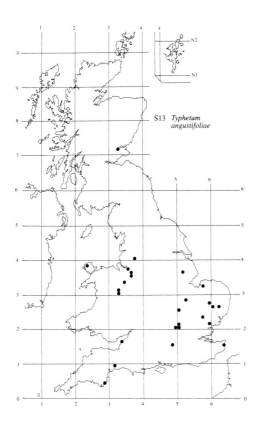

S13 *Typhetum angustifoliae*

S14
Sparganium erectum swamp
Sparganietum erecti Roll 1938

Synonymy
Sparganium ramosum zone Walker 1905; *Sparganium erectum* society Spence 1964; *Sparganium erectum* reedswamp *auct. angl.*

Constant species
Sparganium erectum.

Rare species
Butomus umbellatus, Wolffia arrhiza.

Physiognomy
The *Sparganietum erecti* is generally dominated by *Sparganium erectum* which forms an open or closed cover of shoots about 1 m tall. Although pure and denser stands occur, there are usually some associates and certain of these can attain local prominence. The diversity and moderate species-richness of the vegetation are partly a reflection of the characteristic occurrence of the community in narrow open-water transitions where zonations are often contracted into jumbled mixtures but there are also some morphological and physiological features of the dominant which perhaps permit ready colonisation by other species. The underground organs of *S. erectum* comprises monopodial shoot-producing corms which, though bulky, are relatively short-lived (Cook 1961) and rhizomes which, though sometimes extensive, are much narrower than those of some other swamp dominants (Walker 1905, Cook 1961). The rather open and shifting network of buried organs and aerial shoots may thus leave patches of open water or substrate for invasion. Furthermore, although *S. erectum* thrives best in full sunlight (Cook 1961), it is shade-tolerant and will stand overtopping by other species, provided these are not too bulky. Indeed, elements of this community comprise one of the commonest synusial components of swamps under a variety of other dominants.

Sub-communities

***Sparganium erectum* sub-commuity.** Here are included pure and very species-poor stands in which *S. erectum* is overwhelmingly dominant. There are sometimes floating or floating-leaved aquatics, notably duckweeds, on small areas of open water between the *S. erectum* shoots and *Butomus umbellatus* has been recorded in this vegetation.

***Alisma plantago-aquatica* sub-community.** There may also be a patchy aquatic element in the more species-rich vegetation of this sub-community but the distinguishing feature is the occurrence, beneath and sometimes fronting the *S. erectum*, of an open and fragmentary understorey of species typical of shallow water margins, most frequently *Alisma plantago-aquatica*, sometimes with *Callitriche stagnalis, Nasturtium officinale* or *Apium nodiflorum*. Other swamp dominants, notably *Typha angustifolia*, and tall herbs such as *Epilobium hirsutum* or *Urtica dioica* may occur but these attain no more than local prominence, scattered plants or small clumps protruding above the bur-reed cover.

***Mentha aquatica* sub-community.** In this richest type of *Sparganietum*, the cover of the dominant is generally open and there is an almost complete understorey of a wide variety of associates including small emergent herbs such as *Mentha aquatica, Myosotis scorpioides* and *M. laxa* ssp. *caespitosa* and tussocky monocotyledons, most commonly *Juncus effusus* and *Carex otrubae*. Tall herbs, such as *Lycopus europaeus, Filipendula ulmaria, Iris pseudacorus, Epilobium hirsutum* and *Angelica sylvestris* may break the canopy and there are sometimes small clumps of other swamp dominants, notably *Typha latifolia, T. angustifolia* and *Carex acutiformis*. *Solanum dulcamara* and *Galium palustre* may be prominent as sprawlers and *Holcus lanatus* is frequent though never abundant.

Phalaris arundinacea **sub-community.** Although most of the characterising species of the *Alisma* and *Mentha* sub-communities are absent here, there are occasional tus-socks of *Juncus effusus*, scattered *A. plantago-aquatica* and sometimes floating duckweeds. In general, however, this vegetation is species-poor and distinguished by the consistent presence of small amounts of *Phalaris arundi-nacea*, whose clumps may overtop the bur-reed. *Poly-gonum persicaria* and *Veronica beccabunga* occur occasionally.

Habitat
The *Sparganietum* is a community of shallow, meso-trophic to eutrophic waters with mineral substrates. It occurs widely in the standing waters of small pools, agricultural ponds, dykes and canals, but its tolerance of moderate currents makes it also one of the commonest vegetation types along lowland streams and rivers.

Although *S. erectum* can survive in up to about 1 m of water (Cook 1961) and can even grow when fully submerged (Haslam 1978), the community thrives best in shallows and on water margins but it will not tolerate long periods with a water-table below the roots, about 10 cm below the substrate surface (Cook 1961). To some extent, the floristic variation in the sub-communities is related to the mean water depth, the *Sparganium*, *Alisma*, *Mentha* and *Phalaris* types occurring in progres-sively shallower waters with the last two extending on to the ground which is exposed in summer but winter-flooded.

S. erectum seems to grow best in waters where there is negligible flow and the densest and most luxuriant stands occur in the standing waters of small ponds and narrower dykes. Its roots, however, extend quite deeply in soft substrates and Butcher (1933) encountered the community widely in waters with a flow of up to about 450 m h^{-1}. Scour and spate, though, may damage *S. erectum* by eroding away the substrate from the shallow rhizomes and pulling at the tough leaves, so uprooting the plants (Haslam 1978). Along faster water-courses, therefore, the community often occurs as a fragmentary cover or tends to be confined to havens or the outer curves of bends.

The community occurs over a wide variety of mineral substrates and even occasionally on peat, but it is most characteristic of fine-grained silts and clays which may be anaerobic (Cook 1961).

S. erectum is tolerant of pollution by sewage and some industrial effluents (Haslam 1978) but it will not survive repeated summer cutting or heavy grazing by stock to which it is palatable (Cook 1961).

Zonation and succession
Extensive stands of the community are rare, although the *Sparganium* sub-community sometimes chokes the

centre of small ponds and narrow channels. More often, one or more of the sub-communities occurs in a belt with aquatic communities, other swamps and tall-herb vege-tation in zonations related to water depth and water movement in open-water transitions and as riparian sequences. In the standing waters of ponds and dykes, bands of the *Sparganium* or *Alisma* sub-communities often give way in deeper water to *Potamogeton natans* and duckweeds (e.g. Walker 1905). On gently shelving margins with silty substrates, the *Sparganietum* may form a mosaic with the *Eleocharis* sub-community of the *Eleocharitetum palustris* or grade to it in shallower water. Limits of inundation in fluctuating waters are sometimes marked by a fairly abrupt change to the *Phalaridetum arundinaceae*, and the *Phalaris* sub-community may form a narrow transition. Where graz-ing stock have access and the substrate becomes poached (a common occurrence around streams and farm ponds), a fretted zone of the *Sparganietum* may give way to a patchy *Glycerietum fluitantis* or the *Agros-tis stolonifera-Alopecurus geniculatus* community. Where there is a switch to clayey ground-water gleys on pond margins, there can be a transition through the *Mentha* sub-community of the *Sparganietum* to Holco-Juncion rush-pastures.

Along streams and rivers, the movement of water and the generally more steeply shelving banks often rise to rather different zonations which may be much frag-mented or conflated. A common pattern is for a zone of the *Alisma* sub-community to pass in deeper water to the *Sparganium* sub-communities of the *Typhetum latifoliae* or *Scirpetum lacustris* and to give way in the shallows to the *Alisma-Sparganium* sub-community of the *Glycerie-tum maximae* (e.g. Haslam 1978). On the lower reaches of permanent chalk streams, where clay begins to consti-tute a considerable proportion of the substrate, the *Sparganietum* may occur in similar sequences in a patchy mosaic with the *Caricetum acutiformis* and, along peat dykes, the community sometimes fronts a zone of the *Phragmitetum australis* (e.g. Haslam 1978). In many cases, however, the *Sparganietum* occurs in patchy mosaics along stream and dyke sides with scat-tered clumps of the above communities or with the *Caricetum otrubae* or *Carex pseudocyperus* swamp.

Although *S. erectum* is capable of considerable small rhizome extension and spread of the community may aid the colonisation of small open water bodies, the species does not seem to be able to compete with the more robust *P. australis* and *Typha* spp. (Cook 1961). In running waters any advance of the community is often repeatedly set back by erosion of the banks.

Distribution
The *Sparganietum erecti* is a very common community throughout the agricultural lowlands of England, Wales

and southern Scotland, especially over argillaceous bed-rocks and drift. The distribution of *S. erectum* may be limited northwards and at higher altitudes by low sum-mer temperatures rather than by an inability to tolerate more oligotrophic waters (Spence 1964).

Affinities

Although elements of the community are widely repre-sented as understoreys beneath other swamp dominants characteristic of sluggish, silty, mesotrophic waters, it seems sensible to retain a distinct *Sparganietum erecti* for those very common occurrences where *S. erectum* is the dominant.

Floristic table S14

	a	b	c	d	14
Sparganium erectum	V (5–10)	V (5–10)	V (5–8)	V (6–10)	V (5–10)
Butomus umbellatus	I (5–8)				I (5–8)
Eleocharis palustris	I (2)				I (2)
Alisma plantago-aquatica		IV (1–5)	III (3–4)	II (1)	III (1–5)
Callitriche stagnalis		II (3–8)			I (3–8)
Nasturtium officinale		II (1–4)			I (1–4)
Apium nodiflorum		II (1–4)	I (4)		I (1–4)
Mentha aquatica		I (2)	V (3–8)		II (2–8)
Juncus effusus		I (3)	III (3–4)	II (2–4)	II (2–4)
Myosotis scorpioides		I (3)	III (3–4)		I (3–4)
Lycopus europaeus		I (1–2)	III (3–4)	I (3)	I (1–4)
Solanum dulcamara		I (3–4)	III (2–4)		I (2–4)
Typha latifolia		I (4)	III (3–5)		I (3–5)
Iris pseudacorus		I (3)	III (3–4)		I (3–4)
Filipendula ulmaria			III (1–3)		I (1–3)
Galium palustre			III (3–5)		I (3–5)
Holcus lanatus			III (2–3)		I (2–3)
Carex otrubae			II (3–6)		I (3–6)
Carex acutiformis			II (4–6)		I (4–6)
Juncus articulatus			II (3–4)		I (3–4)
Oenanthe crocata			II (3–4)		I (3–4)
Myosotis laxa caespitosa			II (2–3)		I (2–3)
Ranunculus flammula			II (2–3)		I (2–3)
Scutellaria galericulata			II (1–3)		I (1–3)
Lotus uliginosus			II (4)		I (4)
Angelica sylvestris			II (3)		I (3)
Phalaris arundinacea			I (4)	V (2–4)	II (2–4)
Veronica beccabunga				II (3–4)	I (3–4)
Holcus mollis				II (1–2)	I (1–2)
Apium graveolens				I (5)	I (5)
Agrostis canina canina				I (3)	I (3)

Floristic table S14 (*cont.*)

	a	b	c	d	14
Rumex crispus				I (4)	I (4)
Polygonum persicaria				I (4)	I (4)
Lythrum salicaria				I (3)	I (3)
Hesperis matronalis				I (2)	I (2)
Hydrocharis morsus-ranae				I (5)	I (5)
Nuphar lutea				I (5)	I (5)
Groenlandia densa				I (3)	I (3)
Lemna minor	I (3–5)	II (1–6)		II (1–9)	II (1–9)
Lemna trisulca	I (3–6)	I (9)			I (3–9)
Lemna gibba	I (2)	I (4–6)			I (2–6)
Potamogeton natans	I (3–6)	II (1–9)			I (1–9)
Wolffia arrhiza	I (2–4)			I (2–4)	I (2–4)
Epilobium hirsutum		I (1–5)	II (2–6)	II (1)	I (1–6)
Polygonum amphibium		I (3–5)	I (3)	II (1–4)	I (1–5)
Typha angustifolia		II (3–5)	II (3–6)		I (3–6)
Glyceria fluitans		I (2)	I (3)		I (2–3)
Juncus inflexus		I (4)	I (4)		I (4)
Phragmites australis		I (3)	I (3)		I (3)
Urtica dioica		I (1–3)	I (3)		I (1–3)
Number of samples	15	20	8	8	51
Number of species/sample	2 (1–5)	5 (2–11)	14 (8–24)	6 (2–8)	6 (1–24)
Vegetation height (cm)	99 (45–150)	101 (60–130)	112 (40–150)	108 (80–130)	105 (40–150)
Vegetation cover (%)	84 (25–100)	85 (30–100)	100	90 (70–100)	90 (25–100)

a   *Sparganium erectum* sub-community

b   *Alisma plantago-aquatica* sub-community

c   *Mentha aquatica* sub-community

d   *Phalaris arundinacea* sub-community

14   *Sparganietum erecti* (total)

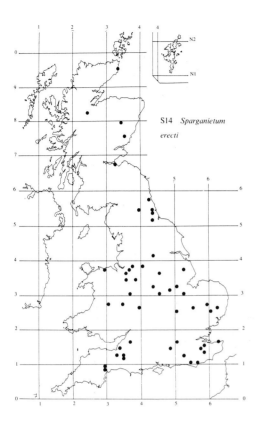

S14   *Sparganietum erecti*

S15
Acorus calamus swamp
Acoretum calami Schulz 1941

Synonymy

Scirpeto-Phragmitetum medioeuropeaum (Koch 1926) R.Tx. 1941 *p.p.*; Gemeenschap van *Acorus calamus* en *Iris pseudacorus* Olivier & Segal 1963 *p.p.*.

Constant species

Acorus calamus.

Rare species

Crassula helmsii.

Physiognomy

The *Acoretum calami* comprises stands dominated by *Acorus calamus* forming an open or closed cover of leafy shoots up to about 1 m tall.

Sub-communities

Acorus calamus **sub-community.** Here are included pure or very species-poor stands of dense *Acorus*.

Lemna minor **sub-community.** In this sub-community, a generally shorter and more open cover of the dominant emerges from open water with a sometimes abundant mat of floating *Lemna minor*. Other swamp dominants such as *Typha latifolia* and *Sparganium erectum* may attain local prominence and there are sometimes tussocks of *Juncus effusus* and sprawls of *Solanum dulcamara* and *Galium palustre*. Tall herbs like *Angelica sylvestris*, *Lycopus europaeus* and *Rumex hydrolapathum* occur occasionally and a wide variety of other water margin species are represented rarely. The introduced Australasian water-plant *Crassula helmsii* (e.g. Clement 1979) has been recorded here.

Habitat

The community occurs in standing or slow-moving waters, 20–80 cm deep, or more rarely on the water-logged margins, of ornamental pools, ponds, canals and dykes. The substrate is usually silt or clay and pH values of 5.4–7.2 have been recorded.

Zonation and succession

Acorus was introduced into Britain, probably before 1660 (Salisbury 1964) and for the medicinal value of its underground organs (Haslam 1978). It was planted thereafter as an ornamental and its sweet-smelling leaves have been used for strewing in churches (Salisbury 1964, Ellis 1965). Isolated stands in ornamental pools often represent deliberate plants but, although seed is not produced in Europe, *Acorus* spreads well by vegetative means and the *Lemna* sub-community perhaps includes stands of natural swamp and marginal vegetation which have been invaded.

Distribution

Acorus occurs in scattered localities throughout the English lowlands with concentrations in Lancashire, the central Midlands and the London Basin.

Affinities

Swamp dominated by *Acorus* has hardly been mentioned in accounts of British vegetation, although a similar community has been described from The Netherlands (Westhoff & den Held 1969). The ecological amplitude of the species in Europe places the *Acoretum* alongside shallow-water swamps of mesotrophic mineral substrates such as the *Sparganietum erecti* and *Typhetum latifoliae*.

S19
Eleocharis palustris swamp
Eleocharitetum palustris   Schennikow 1919

Synonymy
Eleocharis palustris consocies   Pearsall 1918.

Constant species
Eleocharis palustris.

Physiognomy
The *Eleocharitetum palustris* is dominated by an open or closed cover of the slender shoots of *Eleocharis palustris*. There are no other species frequent throughout.

Sub-communities

Eleocharis palustris **sub-community:** *Eleocharis palustris* Community   Birse 1980. Here are included pure and species-poor stands in which *E. palustris* is overwhelmingly the most abundant species. *Alisma plantago-aquatica, Mentha aquatica, Myosotis laxa* ssp. *caespitosa* and *Ranunculus flammula* occur occasionally and a variety of other water-margin species at very low frequency.

Littorella uniflora **sub-community:** *Eleocharis palustris-Littorella* sociation   Spence 1964. In this sub-community *Littorella uniflora* forms a diminutive understorey to the *E. palustris*, sometimes with a little *Lobelia dortmanna*. There is frequently some emergent *Equisetum fluviatile* and floating *Juncus bulbosus* and aquatics such as *Potamogeton natans* and *Myriophyllum alterniflorum* occur occasionally.

Agrostis stolonifera **sub-community:** *Eleocharis palustris-Agrostis stolonifera* nodum   Adam 1981. Here, *Agrostis stolonifera* and, less frequently, *Potentilla anserina*, form a sometimes extensive low mat beneath the *E. palustris* and there are occasional records for species characteristic of the *Juncetum gerardi: Festuca rubra, Juncus gerardi, Glaux maritima, Triglochin maritima* and *Trifolium repens. Eleocharis quinqueflora* may be locally prominent.

Habitat
The *Eleocharitetum* is a swamp of standing or running waters up to 50 cm deep and it occurs round large lakes, small ponds and along stream sides. The sub-communities show some relationships to substrate type and trophic state of the waters. The *Eleocharis* sub-community occurs over various substrates including silts in mesotrophic conditions; the *Littorella* sub-community is more characteristic of sandy or stony substrates in more oligotrophic waters and is common around the shores of some Scottish lakes; the *Agrostis* sub-community occurs in brackish and saline habitats on upper salt-marshes.

Zonation and succession
The community can occur as the distal component of swamp and fen zonations in open-water transitions. The *Littorella* sub-community is commonly found in such situations over sand and gravels around exposed shores in some Scottish lakes (Spence 1964). Here, it grades in deeper water to submerged vegetation with combinations of *L. uniflora, Lobelia dortmanna* and *Juncus bulbosus*. In other cases there is a front of the *Scirpetum lacustris* in deeper water. Inshore, this sub-community can give way to the *Caricetum rostratae*, the *Caricetum vesicariae*, the *Phragmitetum australis* or the *Phalaridetum arudinaceae* (e.g. Pearsall 1918).

In lowland open-water transitions with silty substrates the *Eleocharis* sub-community can again lead the zonation giving way to aquatic vegetation with *Potamogeton natans* and duckweeds in deeper water or occur in shallow water mosaics with the *Equisetetum fluviatile* and the *Sparganietum erecti*.

The *Agrostis* sub-community has been encountered as a dense fringe to streams running across the upper salt-marsh in a few isolated localities in west Wales and around the Scottish coast (Adam 1976, 1981). On Arran, it also occurs in flushes among coastal rocks and here *E. palustris* seems to replace *E. uniglumis* which is very rare on the island but the usual *Eleocharis* sp. dominant in such northern situations (Adam *et al.* 1977).

Distribution

The *Eleocharis* sub-community has a scattered distribution throughout the British lowlands. The *Littorella* sub-community has been recorded only from Scotland, although it probably occurs in the Lake District and Snowdonia. The *Agrostis* sub-community has been encountered in the few localities detailed above.

Affinities

In its various sub-communities, the *Eleocharitetum* shows affinities with various other swamp types, most obviously with the *Equisetetum fluviatile*. Adam (1976, 1981) provisionally assigned his *Eleocharis-Agrostis* nodum to the *Eleocharion uniglumis* along with the *Eleocharetum uniglumis* and the *Blysmetum rufi*. Although the *Agrostis* sub-community shows a similar ecological and geographical distribution to these two communities, it is best kept within this community in a Phragmition or its equivalent.

Floristic table S19

	a	b	c	19
Eleocharis palustris	V (6–10)	V (3–7)	V (7–9)	V (3–10)
Alisma plantago-aquatica	II (1–7)			I (1–7)
Mentha aquatica	II (1–5)			I (1–5)
Myosotis laxa caespitosa	II (1–3)			I (1–3)
Juncus effusus	I (3–5)			I (3–5)
Typha latifolia	I (3–5)			I (3–5)
Phalaris arundinacea	I (1–2)			I (1–2)
Carex hirta	I (2–4)			I (2–4)
Drepanocladus fluitans	I (2–6)			I (2–6)
Juncus acutiflorus	I (3–5)			I (3–5)
Cardamine pratensis	I (2–3)			I (2–3)
Holcus lanatus	I (4)			I (4)
Polygonum persicaria	I (3)			I (3)
Littorella uniflora	I (2)	IV (2–4)		II (2–4)
Equisetum fluviatile	I (2–7)	III (1–3)		II (1–7)
Juncus bulbosus		III (1)		I (1)
Potamogeton natans		II (1)		I (1)
Myriophyllum alternifolium		II (1)		I (1)
Lobelia dortmanna		II (1)		I (1)
Scorpidium scorpioides		I (3)		I (3)
Potamogeton polygonifolius		I (1)		I (1)
Utricularia minor		I (1)		I (1)
Fontinalis antipyretica		I (1)		I (1)
Agrostis stolonifera	I (1–2)		V (2–8)	II (1–8)
Potentilla anserina	I (2–3)		III (2–6)	II (2-6)
Glaux maritima			III (2–5)	I (2–5)
Triglochin maritima			III (2–3)	I (2–3)
Trifolium repens	I (3)		II (5)	I (3–5)
Juncus gerardi			II (4–5)	I (4–5)
Festuca rubra			II (3–5)	I (3–5)
Eleocharis quinqueflora			II (2–4)	I (2–4)
Algal mat			I (6–7)	I (6–7)
Leontodon autumnalis			I (2–4)	I (2–4)
Plantago maritima			I (3)	I (3)

Juncus articulatus	I (3–4)	II (1)	II (3–4)	I (1–4)
Ranunculus acris	I (2–4)	I (1)	I (3)	I (1–4)
Ranunculus flammula	II (1–4)	II (1)		I (1–4)
Polygonum amphibium	I (2–4)	I (1)		I (1–4)
Glyceria fluitans	I (4)	I (1)		I (1–4)
Carex rostrata	I (8)	I (1)		I (1–8)
Caltha palustris	I (1–4)	I (1)		I (1–4)
Hydrocotyle vulgaris	I (4–6)		I (2–4)	I (2–6)
Galium palustre	I (1–3)		I (3)	I (1–3)
Alopecurus geniculatus	I (3–5)		I (2–4)	I (2–5)
Poa trivialis	I (3–4)		I (4)	I (3–4)
Ranunculus repens	I (1–2)		I (5)	I (1–5)
Carex nigra		I (1)	I (4)	I (1–4)
Number of samples	25	11	11	47
Number of species/sample	7 (1–14)			

a   *Eleocharis palustris* sub-community
b   *Littorella uniflora* sub-community
c   *Agrostis stolonifera* sub-community
19 *Eleocharitetum palustris* (total)

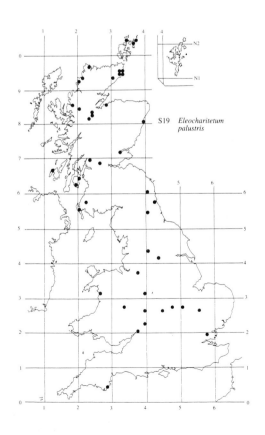

S19   *Eleocharitetum palustris*

S20
Scirpus lacustris ssp. *tabernaemontani* swamp
Scirpetum tabernaemontani　Passarge 1964

Synonymy
Schoenoplectus tabernaemontani nodum　Adam 1976;
Scirpetum maritimae　(Christiansen 1934) R.Tx. 1937
sensu Lee 1977.

Constant species
Scirpus lacustris ssp. *tabernaemontani.*

Physiognomy
In the *Scirpetum tabernaemontani, S. lacustris* ssp. *tabernaemontani* always dominates, its tall shoots forming a dense cover generally 80–90 cm high. *S. maritimus* occurs occasionally, although it rarely exceeds a cover of 25%. Scattered beneath are occasional plants of a variety of salt-marsh species such as *Aster tripolium, Atriplex prostrata, Triglochin maritima, Glaux maritima* and *Cochlearia anglica* and species characteristic of disturbed and/or moist soil surfaces, e.g. *Ranunculus sceleratus, Juncus bufonius* and *Chenopodium rubrum.* There is occasionally a mat of algae over the substrate surface and around the bases of the *Scirpus* stools.

Sub-communities

Sub-community dominated by *Scirpus lacustris* ssp. *tabernaemontani.* Here the dominant occurs alone or with a very sparse understorey of salt-marsh or freshwater swamp associates.

***Agrostis stolonifera* sub-community.** *Agrostis stolonifera* forms an open mat beneath the *Scirpus* and associates are more frequent and abundant. *Eleocharis uniglumis, E. palustris, Carex otrubae* and *Oenanthe lachenalii* are differential occasionals.

Habitat
The community occurs most frequently in moist, brackish sites with soft, anaerobic raw gleys of silt or clay. It may be encountered in depressions at various levels on coastal salt-marshes, around flashes in inland saline sites and as emergent vegetation in counter-dikes, warping-drains and as a fringe around some Scottish sea-lochs.

Although *S. lacustris* ssp. *tabernaemontani* appears to be tolerant of salinities similar to those endured by *S. maritimus* (Packham & Liddle 1970) and will survive surface salt efflorescence (Lee 1977), it is also able to grow in standing fresh water and the *Scirpus*-dominated sub-community of the *Scirpetum tabernaemontani* occurs locally inland.

S. lacustris ssp. *tabernaemontani* is grazed by cattle and Lee (1977) considered that this may play some part in confining the community to wetter, less accessible sites. Large-billed grey geese also eat the shoots as well as digging for roots and rhizomes (Ogilvie 1978).

Zonation and succession
Stands are usually well marked off from the surrounding vegetation by virtue of the bulky physiognomy of the dominant, although scattered shoots of dwarf *S. lacustris* ssp. *tabernaemontani* occur widely in a variety of upper-marsh grasslands.

Distribution
S. lacustris ssp. *tabernaemontani* is an under-recorded taxon in Britain. The community occurs on and close to much of the English and Welsh coasts and more sporadically in Scotland. Inland situations are very local throughout the English lowlands.

Affinities
The vegetation included here is very similar to that of the *Scirpetum maritimi* and stands in which both *Scirpus* taxa occur are fairly frequent. *S. lacustris* ssp. *tabernaemontani* vegetation is sometimes included within the *Scirpetum maritimi* (Westhoff & den Held 1969) or within the *Scirpetum lacustris.*

Floristic table S20

	a	b	20
Scirpus lacustris tabernaemontani	V (6–9)	V (7–10)	V (6–10)
Puccinellia distans	I (3)		I (3)
Typha angustifolia	I (3–5)		I (3–5)
Rumex hydrolapathum	I (1–4)		I (1–4)
Solanum dulcamara	I (1–4)		I (1–4)
Phragmites australis	I (2–7)		I (2–7)
Puccinellia maritima	I (3–4)		I (3–4)
Ruppia maritima	I (5–8)		I (5–8)
Agrostis stolonifera		V (1–8)	II (1–8)
Eleocharis palustris		II (1–4)	I (1–4)
Carex otrubae		II (2–5)	I (2–5)
Oenanthe lachenalii		II (3–4)	I (3–4)
Eleocharis uniglumis		II (3–4)	I (3–4)
Apium graveolens		I (4)	I (4)
Potentilla anserina		I (3–5)	I (3–5)
Scirpus maritimus	II (2–7)	II (3–5)	II (2–7)
Algal mat	II (4–7)	II (6–8)	II (4–8)
Aster tripolium	I (3–7)	II (3–6)	I (3–7)
Atriplex prostrata	I (3)	II (3–7)	I (3–7)
Triglochin maritima	I (2–4)	II (2–5)	I (2–5)
Glaux maritima	I (3)	II (2–5)	I (2–5)
Cochlearia anglica	I (2)	II (2–3)	I (2–3)
Typha latifolia	I (3–4)	I (4)	I (3–4)
Chenopodium rubrum	I (3)	I (3–4)	I (3–4)
Ranunculus sceleratus	I (1)	I (1–2)	I (1–2)
Juncus bufonius	I (4)	I (3)	I (3–4)
Spartina anglica	I (2)	I (2)	I (2)
Juncus gerardi	I (4)	I (3)	I (3–4)
Number of samples	32	15	47
Number of species/sample	3 (1–8)	7 (3–19)	5 (1–19)
Vegetation height (cm)	90 (60–200)	80 (50–150)	87 (50–200)
Vegetation cover (%)	84 (60–100)	86 (70–100)	84 (60–100)

a   *Scirpus lacustris tabernaemontani*-dominated sub-community
b   *Agrostis stolonifera* sub-community
20 *Scirpetum tabernaemontani* (total)

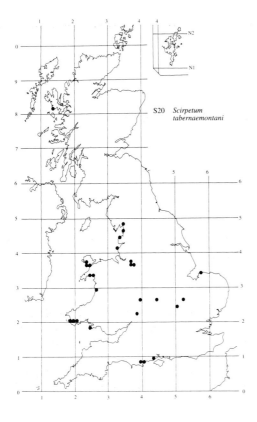

S20 *Scirpetum*
     tabernaemontani

S21

Scirpus maritimus swamp
Scirpetum maritimi (Br.-Bl. 1931) R.Tx. 1937

Synonymy
Scirpus maritimus zone   Willis & Davies 1960; Stands dominated by *Scirpus maritimus* Gimingham 1964; *Scirpetum maritimi* Packham & Liddle 1970; Brackish water communities Birks 1973 *p.p.*; *Scirpus maritimus* nodum   Adam 1976.

Constant species
Scirpus maritimus.

Rare species
Juncus subulatus.

Physiognomy
The *Scirpetum maritimi* is always dominated by *Scirpus maritimus* which usually forms a tall dense cover, about 60–80 cm high in which up to 10 shoots dm^{-2} have been recorded (Gimingham 1964*a*). No other species is frequent throughout. Pure stands are common and very rarely are there more than ten associates. Most often, there are scattered plants of a wide range of species characteristic of the upper salt-marsh and strandline: *Triglochin maritima, Juncus gerardi, Oenanthe lachenalii, Cochlearia anglica, Plantago maritima, Elymus pycnanthus* and *Apium graveolens* are those most commonly encountered. Even in the richer sub-communities, which are distinguished by the constancy of additional species, the associates rarely exceed a combined cover of 30% and most frequently form a patchy cover beneath the canopy of *S. maritimus. S. lacustris* ssp. *tabernaemontani* is rare here and never abundant.

Bryophytes are very sparse but there may be a prominent mat of algae over the substrate surface and around the bases of the *Scirpus* stools. Gimingham (1964*a*) noted the abundance of *Vaucheria* spp. in this algal mat in Scottish stands of the community.

The Mediterranean maritime rush, *Juncus subulatus*, has become naturalised within the *Scirpetum maritimi* on the young salt-marsh at Berrow in Somerset (Willis & Davies 1960). It successfully competes against *S. maritimus* there by healthy rhizome extension.

Sub-communities

Sub-community dominated by *Scirpus maritimus*. In the very species-poor vegetation of this sub-community, *S. maritimus* occurs alone or with one or two associates of salt-marsh communities and freshwater swamps. Surface algae, *Ruppia maritima* or tall swamp species such as *Phragmites australis* or *Glyceria maxima* occasionally attain local prominence.

***Atriplex prostrata* sub-community:** *Halo-Scirpetum maritimi* (van Langendonck 1931) Dahl & Hadač 1941. Here, *Atriplex prostrata* is constant and forms, with the less frequent *Puccinellia maritima*, rayed *Aster tripolium, Triglochin maritima* and a variety of upper-marsh species, an open ground cover.

***Agrostis stolonifera* sub-community.** *A. prostrata* remains constant in this sub-community but the most prominent feature of the vegetation is an open carpet of *Agrostis stolonifera*. Scattered through this, *Triglochin maritima, Glaux maritima, Juncus gerardi* and *Oenanthe lachenalii* occur and each may be abundant in particular stands. In this sub-community and the next there are also rare records for a variety of species common in inland freshwater and disturbed habitats, e.g. *Glyceria fluitans, Caltha palustris, Carex nigra, Viola palustris, Cirsium arvense* and *Stellaria media*.

***Potentilla anserina* sub-community.** The major species of the *Agrostis* sub-community retain their frequency here but, in addition, *Potentilla anserina* is constant, although it is rarely abundant. *Rumex crispus, Cochlearia officinalis, Festuca arundinacea* and *Juncus bufonius* also occur occasionally. This is the richest of the sub-communities and bryophytes may be locally conspicuous with *Drepanocladus aduncus, Amblystegium serpens* and *Calliergon cuspidatum.*

Habitat

The community is characteristic of ill-drained brackish sites on coastal salt-marshes, occuring as often small patches in pans, borrow pits and alongside creeks, usually on the upper marsh, and in estuaries where stands may be more extensive. It has also been recorded from counter-dikes behind reclamation banks and from ditches and flashes in inland saline sites in Cheshire and Lancashire.

The *Scirpetum maritimi* seems to tolerate conditions which are both wetter and more saline than those typically favoured by the *Phragmites-Atriplex* swamp. The soils are characteristically unripened raw gleys of soft silt or clay, black and foul-smelling and sometimes with an orange crust. The water-table may fluctuate: salt-marsh stands are occasionally inundated by tidal waters, estuary stands frequently covered by brackish waters and the community can grow as emergent vegetation in more than 50 cm of standing water. However, a distinctive feature of the habitat seems to be the absence of a constant throughput of water, either brackish or fresh (e.g. Jermy & Crabbe 1978; cf. Chapman 1960*a*). In such situations, *S. maritimus* has been shown to tolerate salinities of up to about 20 g l^{-1} (chloridity 12 g l^{-1}: Willis & Davies 1960, Ranwell *et al.* 1964, Proctor 1980).

More exceptionally, the *Scirpetum maritimi* occurs on patches of rotting seaweed over shingle in western Scotland (Birks 1973) and the vegetation will also tolerate a certain amount of burial in blown sand. The community occurs on both grazed and ungrazed marshes. *Scirpus* shoots, roots and rhizomes are eaten by large-billed grey geese (Ogilvie 1978).

Zonation and succession

The *Scirpetum maritimi* can occur at various levels on coastal salt-marshes and, exceptionally, there may be a zonation within stands between the various sub-communities, from the *Scirpus*-dominated type at the lowest level, through the *Atriplex* and *Agrostis* sub-communities, to the *Potentilla* sub-community. More often, stands of one or other of the sub-communities, sometimes quite small in extent, occur within other salt-marsh communities, forming mosaics with the *Puccinellietum maritimae* or, more usually, the *Juncetum gerardi*, the *Juncetum maritimi* (Packham & Liddle 1970) or the *Festuca rubra-Agrostis stolonifera-Potentilla anserina* mesotrophic grassland. The boundaries of the *Scirpetum maritimi* are generally sharply marked by the dominance of *S. maritimus* but the vegetation types may grade one into another through the associates in the subsidiary layer of the swamp.

In estuaries, the *Scirpetum maritimi* is sometimes the pioneer vegetation in an inverted zonation, as on the Exe for example, where it occurs down-slope from the *Spartinetum townsendii* and the *Juncetum gerardi* (Proctor 1980).

The community also grades to other swamp types in salt-marsh depressions or on estuarine foreshores. It frequently passes gradually into the *Scirpetum tabernaemontani* with a switch in dominance between the two *Scirpus* taxa and mosaics with the *Phragmitetum australis* are common. On the Nith in the Solway Firth and at Dingwall in the Cromarty Firth, the *Scirpetum maritimi* occurs seaward of the *Atriplex* sub-community of this swamp type.

In a number of sheltered sea-lochs in western Scotland, the community occurs as a fringe of emergent vegetation giving way sharply above to *Alnus* woodland, a zonation which is widespread in the Baltic (Tyler 1969*b*).

Distribution

The *Scirpetum maritimi* occurs in suitable sites on all coasts of the British Isles as far north as Sutherland. The *Scirpus*-dominated and *Atriplex* sub-communities have been recorded throughout the range; the *Agrostis* and *Potentilla* sub-communities seem to be more common on the grazed sandy marshes of the west coast.

Affinities

Although the community is readily recognisable, the species-poor nature of much of the vegetation makes it difficult to define its affinities and phytosociological treatments have been confused. Some authorities regard this vegetation as a form of *Scirpo-Phragmitetum*; others give it association status still within the Phragmition alliance. Yet others place a *Scirpetum maritimi* within a completely separate class, the Bulboschoenetea, so as to give prominent recognition to the scarcity of Phragmition and Magnocaricion associates (e.g. Birse 1980). .

On occasion, a distinction has been made within vegetation of this kind between very species-poor stands and those which have prominent halophyte associates. The former, in which the dominant taxon is *S. maritimus* var. *maritimus*, have been retained as a *Scirpetum maritimi*; the latter, distinguished by the dominance of *S. maritimus* var. *compactus*, have been separated off into a *Halo-Scirpetum* and placed in a Halo-Scirpion alliance within the Asteretea (e.g. Westhoff & den Held 1969). Although this view has been widely adopted in Holland, these two *Scirpus* taxa have not generally been distinguished in Britain and are not separated in the most recent European treatment of the genus (DeFilipps 1980). Although the vegetation described here as the *Atriplex* sub-community clearly corresponds to the *Halo-Scirpetum*, it is no more distinctive than the other two richer sub-communities and it seems preferable to retain all within a single vegetation type.

Since its original description by Willis & Davies (1960), *Juncus subulatus* has considerably expanded its cover at Berrow and is less obviously part of the *Scirpetum maritimi* in which it originally gained a hold. Within its chief area of distribution, around the Mediterranean, the Caspian and the north Spanish coast, this species occurs in the perennial glasswort vegetation of the *Salicornietum fruticosae* (Braun-Blanquet 1952), with *Juncus maritimus* (Molinier 1948, Gradstein & Smittenburg 1977) and with *Scirpus maritimus* (Holmboe 1914, Bolos & Bolos 1950). Willis & Davies (1960) noted the interesting parallels between the British occurrence of *J. subulatus* and two other rarities with a foothold in the south-west maritime zone, *Scirpus holoscheonus* and *Juncus mutabilis*.

Floristic table S21

	a	b	c	d	21
Scirpus maritimus	V (6–10)	V (6–10)	V (5–10)	V (7–10)	V (5–10)
Atriplex prostrata		V (1–6)	IV (1–6)	III (2–4)	III (1–6)
Puccinellia maritima	I (2–6)	II (1–7)	I (2–6)	I (3)	I (1–7)
Aster tripolium (rayed)	I (2–5)	II (2–6)	I (2–5)		I (2–6)
Aster tripolium var. *discoideus*		I (2)			I (2)
Agrostis stolonifera			V (3–7)	V (3–6)	II (3–7)
Triglochin maritima	I (3–6)	II (2–4)	III (2–5)	III (2–5)	II (2–6)
Glaux maritima	I (1–4)		II (2–5)	II (4)	I (1–5)
Juncus gerardi	I (2–3)	I (2–3)	II (2–6)	III (2–6)	I (2–6)
Oenanthe lachenalii	I (2–4)	I (2)	II (1–6)	I (3–4)	I (1–6)
Juncus maritimus			I (3–7)		I (3–7)
Puccinellia distans			I (5–8)		I (5–8)
Samolus valerandi			I (1)		I (1)
Potentilla anserina				V (2–7)	I (2–7)
Rumex crispus	I (1–3)	I (1–3)	I (1–6)	II (2–4)	I (1–6)
Cochlearia officinalis			I (1–4)	II (1–2)	I (1–4)
Festuca arundinacea	I (3)	I (2)	I (2–4)	II (2–3)	I (2–4)
Drepanocladus aduncus			I (3)	II (1–4)	I (1–4)
Juncus bufonius			I (4)	II (2)	I (2–4)
Juncus articulatus				I (1–3)	I (1–3)
Amblystegium serpens				I (4)	I (4)
Elymus pycnanthus	I (2–6)	I (3–7)	I (4)	I (3)	I (2–7)
Cochlearia anglica	I (1–4)	I (2–3)	I (2–5)	I (1)	I (1–5)
Apium graveolens	I (3–4)	I (2)	I (3–5)	I (5)	I (2–5)
Carex otrubae	I (3)	I (3)	I (3–4)	I (1)	I (1–4)
Plantago maritima	I (2–3)	I (2–3)	I (3)	I (3)	I (2–3)
Scirpus lacustris tabernaemontani	I (2)	I (2)	I (3–6)	I (3)	I (2–6)
Calystegia sepium	I (2)	I (3)	I (3–4)	I (3)	I (2–4)

Species				
Algal mat	II (4–8)	II (4–8)		I (4–8)
Spartina anglica	I (2–3)	I (2)		I (1–3)
Suaeda maritima	I (3–6)	I (3)		I (3–6)
Aster tripolium	I (2–3)	I (1–3)		I (1–3)
Halimione portulacoides	I (3–4)	I (2)		I (2–4)
Spergularia marina	I (1)	I (3–4)		I (1–7)
Sonchus arvensis	I (2)	I (3)		I (1–4)
Berula erecta	I (3–4)	I (3–5)		I (3–5)
Phragmites australis	I (2–5)	I (3–4)		I (2–5)
Mentha aquatica	I (3)	I (3–6)		I (3–6)
Limonium cf. vulgare	I (2)	I (1)		I (1–2)
Salicornia dolichostachya	I (2–3)	I (3)		I (2–3)
Epilobium parviflorum	I (3)	I (2)		I (2–3)
Glyceria maxima	I (5)	I (4)		I (4–5)
Ruppia maritima	I (8)	I (5)		I (5–8)
Spergularia media	I (3)	I (2)		I (2–3)
Iris pseudacorus	I (1–2)	I (2–3)	I (1)	I (1–3)
Galium palustre	I (4)	I (2–4)	I (1)	I (1–4)
Hydrocotyle vulgaris	I (3)	I (2–9)	I (4)	I (2–9)
Ranunculus sceleratus	I (1)	I (2–3)	I (2)	I (1–3)
Tripleurospermum maritimum	I (2–3)	I (2)	I (2)	I (2–3)
Matricaria maritima	I (2–3)	I (2)	I (2)	I (2–3)
Elymus repens	I (3)	I (2–6)	I (3)	I (2–6)
Galium aparine	I (2)	I (1–2)		I (1–2)
Oenanthe crocata	I (2)	I (2–5)		I (2–5)
Rumex conglomeratus	I (3)	I (2)		I (2–3)
Eleocharis palustris		I (3–5)	I (1–2)	I (1–5)
Eleocharis uniglumis		I (4–6)	I (4)	I (4–6)
Caltha palustris		I (4)	I (2–3)	I (2–4)
Festuca rubra		I (4)	I (4)	I (4)
Glyceria fluitans		I (2)	I (3–4)	I (2–4)
Carex nigra		I (5)	I (3)	I (3–5)
Cirsium arvense		I (3)	I (2)	I (2–3)
Stellaria media		I (4)	I (4)	I (4)
Viola palustris		I (2)	I (2)	I (2)

Floristic table S21 (*cont.*)

	a	b	c	d	21
Calliergon cuspidatum			I (5)	I (2)	I (2–5)
Ranunculus flammula			I (3)	I (3)	I (3)
Number of samples	54	28	37	12	131
Number of species/sample	3 (1–12)	4 (2–9)	7 (3–13)	10 (6–20)	5 (1–20)
Vegetation height (cm)	81 (37–150)	78 (40–120)	69 (25–130)	73 (35–120)	76 (25–150)
Vegetation cover (%)	87 (40–100)	95 (80–100)	93 (60–100)	91 (75–100)	91 (40–100)

a *Scirpus maritimus*-dominated sub-community
b *Atriplex prostrata* sub-community
c *Agrostis stolonifera* sub-community
d *Potentilla anserina* sub-community
21 *Scirpetum maritimi* (total)

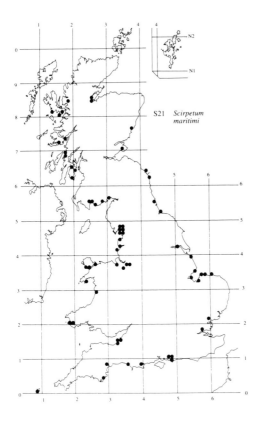

S21   *Scirpetum maritimi*

S22
Glyceria fluitans water-margin vegetation
Glycerietum fluitantis   Wilczek 1935

Constant species

Glyceria fluitans.

Physiognomy

The *Glycerietum fluitantis* is dominated by a low mat or floating carpet of *Glyceria fluitans*, sometimes continuous and very species-poor, in other cases as a more open cover with a variety of associates, some of which may attain local prominence. No other species reaches even occasional frequency throughout but the most usual associates are plants of shallow water margins such as *Alisma plantago-aquatica*, *Myosotis scorpioides*, *Apium nodiflorum* and *Eleocharis palustris*.

Sub-communities

Glyceria fluitans sub-community. This sub-community includes pure or very species-poor stands in which *G. fluitans* forms an often thick and lush mat. *Lemna minor* is the only frequent associate.

Sparganium erectum-Mentha aquatica sub-community. *Sparganio-Glycerietum fluitantis* Br.Bl. 1925. Here, there is a usually more open cover of the dominant among clumps of *Typha latifolia*, *Sparganium erectum*, *Nasturtium officinale* and *Carex otrubae* with scattered plants of *Mentha aquatica*, *Myosotis scorpioides*, *Alisma plantago-aquatica* and *Rumex crispus*.

Alopecurus geniculatus sub-community: Washlands and wet alluvial meadows   Ratcliffe 1977 *p.p.. Alopecurus geniculatus* is often abundant and sometimes co-dominant in this sub-community and there is occasionally a little *Poa trivialis*. *Rumex crispus* is preferentially frequent here.

Habitat

The community is characteristic of shallow, standing or sluggish, mesotrophic waters and fine mineral substrates, and is commonly found around ponds and wet depressions in fens and pastures and on the margins of dykes and small streams. The *Glyceria* sub-community includes stands where a quaking mass of *G. fluitans* extends out into small areas of sometimes deeper open water in fens or in other undisturbed situations where the dominant can pre-empt a niche and spread. The *Alopecurus* sub-community is more restricted to shallower waters and occurs around the gently sloping silty margins of ponds and pasture hollows. The *Sparganium-Mentha* sub-community is an often narrow and fragmentary community where *G. fluitans* gains a hold among more robust swamp species on stream sides and around pond edges.

G. fluitans is a succulent and palatable grass which is avidly eaten by stock and herbivorous wildfowl. On the Fen washlands, this community formed part of the swamp/grassland mosaics traditionally maintained by winter flooding and summer grazing and now surviving largely on the Ouse Washes where they provide valuable grazing for a variety of bird populations. Poaching by stock can break up the cover of the community and perhaps aid its spread by the trampling in of broken stems which readily re-root in moist ground.

Zonation and succession

Sometimes, the three sub-communities occur as a zonation around the edges of deeper ponds. Often, however, small stands of the *Glycerietum* form a mosaic among other swamp communities on the more open margins of pools and streams. In wet pasture hollows, the *Alopecurus* sub-community may give way on drier ground to the *Lolio-Cynosuretum* and on coarser inundated sediments on river banks to Elymo-Rumicion vegetation.

On the Ouse Washes the community occurs in mosaics with the *Agrostis stolonifera-Alopecurus geniculatus* inundation grassland in areas which are sheep-grazed in summer and subject to shallow winter flooding.

Distribution
The *Glycerietum* is widespread and common in the agricultural lowlands.

Affinities
The community shows most similarities with those swamps where, along shallow and sometimes disturbed water margins, the integrity of the synusial element becomes fragmented and a variety of different species may attain local dominance, e.g. the *Caricetum otrubae* and *Carex pseudocyperus* swamps and vegetation with *Apium nodiflorum* and *Nasturtium officinale*. Such vegetation has sometimes been placed in the Glycerio-Sparganion, in other cases in the Elymo-Rumicion.

Floristic table S22

	a	b	c	22
Glyceria fluitans	V (6–10)	V (6–8)	V (6–10)	V (6–10)
Lemna minor	III (2–7)			II (2–7)
Elodea canadensis	I (1–6)			I (1–6)
Typha latifolia	I (1–4)	III (2–4)	I (3)	II (1–4)
Nasturtium officinale	I (1–2)	III (1–4)		I (1–4)
Sparganium erectum		III (1–5)		I (1–5)
Mentha aquatica		III (4–5)		I (4–5)
Alisma plantago-aquatica	I (2–3)	II (2–3)	I (2)	I (2–5)
Myosotis scorpioides	I (1–4)	II (3)	I (1)	I (1–4)
Juncus articulatus		II (3)	I (3)	I (3)
Carex otrubae		II (3–4)		I (3–4)
Ranunculus lingua		I (4)		I (4)
Solanum dulcamara		I (3)		I (3)
Alopecurus geniculatus	I (4)		V (3–8)	II (3–8)
Poa trivialis		I (4)	II (2–3)	I (2–4)
Rumex crispus	I (3)	II (1–3)	III (2–3)	I (1–3)
Galium palustre	I (2)	I (3)	I (2)	I (2–3)
Agrostis stolonifera	I (1–5)	I (4)		I (1–5)
Apium nodiflorum	I (4)		I (4)	I (4)
Eleocharis palustris	I (3)		I (3)	I (3)
Number of samples	27	7	7	41
Number of species/sample	3 (1–8)	10 (5–22)	6 (4–10)	5 (1–22)

a   *Glyceria fluitans* sub-community
b   *Sparganium erectum-Mentha aquatica* sub-community
c   *Alopecurus geniculatus* sub-community
22 *Glycerietum fluitantis* (total)

S23
Other water-margin vegetation
Glycerio-Sparganion Br.-Bl. & Sissingh *apud* Boer 1942 *emend.* Segal

Constant species

The vegetation included here is characteristically heterogeneous but the most frequent species are *Apium nodiflorum*, *Nasturtium officinale* (including *N. microphyllum*) and *Veronica beccabunga*. Some authorities have also erected vegetation types on the constancy of *Catabrosa aquatica* or *Glyceria plicata*.

Physiognomy

Though floristically varied, this vegetation has a highly distinctive structure. Most of the species are small perennials with shoots that are procumbent and rooted below and ascending (or sometimes floating) above. These form a bushy and often patchy canopy of somewhat uneven height, though generally less than 30 cm. Each of the most frequent species can be locally dominant, forming quite large pure clumps which monopolise long stretches of water margin or alternate with patches of other species. In other cases, there is a more intimate mixture of smaller individuals of two or more of these most frequent species with a variety of associates, the commonest of which are *Myosotis scorpioides*, *Mentha aquatica*, *Alisma plantago-aquatica*, *Berula erecta* (in the south and east) and small *Glyceria* spp. (*G. fluitans*, *G. plicata*, their hybrid *G.* × *pedicellata* or *G. declinata*). Each of these may also assume local prominence, too, and there may be patches of *Ranunculus sceleratus*, *Juncus bufonius* or *Callitriche stagnalis* on bare mud. Then, there are often occasional records for species from a variety of vegetation types which are frequently in close juxtaposition at water margins, most notably swamps, tall-herb fens and some mesotrophic grasslands. Added to this spatial variation, there is often an element of seasonal change in the vegetation in relation to natural periodicity in growth patterns (*Apium nodiflorum*, for example, often spreading markedly in late summer and *Berula erecta* in early spring: Haslam 1978) or in response to disturbance such as dredging or cutting.

Habitat

This vegetation is most typical of the unshaded margins of mesotrophic to eutrophic waters where there is some accumulation of medium- to fine-textured mineral sediments. It may be a temporary feature fronting emergents as they colonise shallow open waters but, where there is some measure of marginal disturbance by turbulence in running waters or from dredging, cutting or moderate trampling or where there is seasonal drying up of standing or running waters, it often forms a permanent, though temporally variable, fringe. It is very common around small standing waters like field ponds, along canal and dyke margins and in the shallows of low-order streams.

The most frequent and abundant species are well adapted to the moderate or periodic instability in such habitats. Though they offer a high resistance to moving waters because of their bushy habit and are often shallow-rooted (both features in which they differ markedly from most emergents), they are relatively robust and can survive in moderate to fast flows. Spates tend to uproot rather than break the plants (Haslam 1978) but they can readily re-root if thrown up in congenial situations. Of all the communities in this section (except perhaps the *Phalaridetum*), it is this kind of emergent vegetation which penetrates furthest towards the head of rivers and, over a wide variety of bedrocks, it constitutes the bulk of any emergent cover along the margins of swifter streams.

In such situations, these species are well able to encourage the dropping of suspended sediments by interrupting the flow of the moving waters and trapping particulate matter among their numerous branching shoots. Most can also readily produce nodal roots at progressively higher levels, thus surviving any deposition, and a capacity for rapid growth enables them to capitalise on even temporary sedimentation.

It is features such as these which also enable this vegetation to respond quickly after artificial disturbance and these species are often the first to return along dyke and canal margins after dredging and cutting.

Small shifts in the level of the water-table do not seem to affect the vegetation adversely, even when they are unseasonal. The shoots of many species can float in shallow water and some have thinner cuticular layers on their lower organs, which suggests that they may be able to benefit from surface exchange of dissolved gases and nutrients. Indeed, on occasion, some of the species, e.g. *Apium nodiflorum* and *Berula erecta*, are encountered as submerged carpets. The vegetation can also tolerate the temporary drying out of shallow waters such as occurs in small ponds and in the upper courses of Chalk winterbournes.

Zonation and succession

In shallow, standing or sluggish waters, the vegetation often occurs as a fringe to the *Glycerietum maximae*, the *Typhetum latifoliae* and the *Sparganietum erecti* and in more intact sequences, some of its species may run under more open stands of these emergents to constitute a sparse understorey of the assemblage G type. Often, however, zonations are very much compressed and, along many dykes and canals, clumps of this vegetation alternate with stands of emergents. These more patchy and fragmentary mixtures also occur with swamps like the *Caricetum acutiformis*, the *Acoretum calami* and the *Sagittaria sagittifolia* and *Carex pseudocyperus* communities.

In running waters with a moderate flow, the most frequent neighbouring swamp is the *Sparganietum erecti* but, in more swiftly moving waters, even this may disappear, leaving a broken fringe of this Glycerio-Sparganion vegetation, perhaps backed by clumps of the *Phalaridetum*. Such patterns are also characteristic of the margins of some more permanent river shoals.

Another very common pattern, around the edges of ponds and along stream margins where stock water, is for this vegetation to occur with the *Glycerietum fluitan-tis* in transitions to damp mesotrophic pastures such as the *Holco-Juncetum* or certain kinds of *Lolio-Cynosuretum*. Slumped fragments of these communities often occur intermixed with Glycerio-Sparganion vegetation in a confusing jumble along stream beds.

Periodic disturbance often sets back any seral progression in the habitats in which this vegetation typically occurs but, in more stable situations, it seems to give way to swamps like the *Glycerietum maximae*, the *Typhetum latifoliae* and the *Sparganietum erecti*. Its most frequent species are light-demanding (Haslam 1978) and, though they may persist for some time as a swamp understorey, they disappear as the emergents advance and become more dense.

Distribution

This vegetation is very common throughout the lowlands of Britain though certain species are more frequent in some regions than others: *Berula erecta*, for example, is largely restricted to the south and east and *Apium nodiflorum* seems particularly common in the south-west. There are also some associations with geological variations: *Nasturtium officinale* is especially prominent in Chalk streams and *Veronica beccabunga* seems better able than many of the other species to survive in streams over nutrient-poor sedimentary rocks.

Affinities

The general affinities of this vegetation are clearly with the Glycerio-Sparganion alliance. With further sampling, it would probably be possible to recognise distinct communities dominated by *Nasturtium officinale* (as in Seibert 1962, Oberdorfer 1977), *Apium nodiflorum* (cf. Westhoff & den Held 1969) and *Veronica beccabunga* (the *V. beccabunga*-Gesellschaft of Oberdorfer 1977). Birse (1980) recognised a *Catabrosetum aquaticae* Rübel 1912 and some Continental studies have characterised a *Glycerietum plicatae* (Kulcz. 1928) Oberdorfer 1954.

S24

Phragmites australis-Peucedanum palustre tall-herb fen
Peucedano-Phragmitetum australis Wheeler 1978 emend.

Synonymy

Fen Association Pallis 1911; *Junco subnodulosi-Calamagrostietum canescentis* Korneck 1963 *p.p.*; *Cladietum marisci sensu* Krausch 1964 *p.p.* and *sensu* Westhoff & den Held 1969 *p.p.*; *Scirpo-Phragmitetum sensu* Krausch 1964 *p.p.* and *sensu* Westhoff & den Held 1969 *p.p*; *Thelypterido-Phragmitetum* Westhoff & den Held 1969 *p.p.*; *Thelypterido-Phragmitetum* Kuiper 1957 *emend.* Segal & Westhoff 1969 *p.p.*; *Peucedano-Phragmitetum australis* Wheeler 1978 *p.p.* and *Caricetum paniculatae peucedanetosum* Wheeler 1978.

Constant species

Calamagrostis canescens, Cladium mariscus, Eupatorium cannabinum, Filipendula ulmaria, Galium palustre, Juncus subnodulosus, Lysimachia vulgaris, Lythrum salicaria, Mentha aquatica, Peucedanum palustre, Phragmites australis, Calliergon cuspidatum.

Rare species

Carex diandra, C. appropinquata, Cicuta virosa, Lathyrus palustris, Peucedanum palustre, Sium latifolium, Thelypteris palustris.

Physiognomy

The *Peucedano-Phragmitetum* is a community of herbaceous fen vegetation, somewhat varied in composition but often species-rich and of complex physiognomy (Wheeler 1975, 1978, 1980a). Tall monocotyledons generally make up the major structural component and, of these, *Phragmites australis, Cladium mariscus* and *Calamagrostis canescens* are constant and the most usual dominants. Much less frequent throughout but occasionally dominant are *Carex paniculata, Glyceria maxima, Typha angustifolia, Phalaris arundinacea* and *Calamagrostis epigejos*. The gross appearance of the vegetation at any particular time of year depends very much on the phenology of these species and on the fen treatment, if any, but in summer, unmown stands have a

characteristically tall canopy, 1–2 m high, of varying density but often sufficiently thick to give stands a superficially uniform appearance.

Intermixed with these helophytes is a variety of tall herbaceous dicotyledons which, though rarely individually abundant, together give the vegetation a typically colourful appearance in the flowering season. *Lysimachia vulgaris, Lythrum salicaria, Eupatorium cannabinum, Filipendula ulmaria* and *Peucedanum palustre* are constant, the rare milk parsley having its major locus in this community. Also frequent are *Valeriana officinalis, Iris pseudacorus* and *Lycopus europaeus* and, rather more unevenly between the various sub-communities, *Angelica sylvestris, Cirsium palustre, Rumex hydrolapathum, Epilobium hirsutum* and *Symphytum officinale.*

Beneath these components, there is a rather variable layer, some 60–80 cm tall, generally dominated by sedges or rushes. *Juncus subnodulosus* and *Carex elata* are the most frequent and abundant species here but *C. riparia, C. acutiformis, C. appropinquata, C. lasiocarpa* or *C. diandra* can all attain local prominence. In other cases, *Molinia caerulea, Schoenus nigricans* or *Thelypteris palustris* are abundant in this layer.

Smaller herbaceous species show a somewhat patchy occurrence throughout the community, their variety and abundance being greater where the layers of taller and bulkier species are more open. *Mentha aquatica* is the only frequent species throughout but some of the following often occur: *Scutellaria galericulata, Hydrocotyle vulgaris, Potentilla palustris, Oenanthe fistulosa, Epilobium palustre, Cardamine pratensis, Lychnis floscuculi* and *Myosotis laxa* ssp. *caespitosa.* In some of the wetter expressions of the community, floating-leaved aquatics may be present.

A variety of sprawlers and climbers may be found, sometimes binding the vegetation into an almost impenetrable tangle. The most frequent of these is *Galium palustre* but *Solanum dulcamara, Calystegia sepium, Vicia cracca, Galium uliginosum* and the rarer *Stellaria palustris* and *Lathyrus palustris* also occur.

*lliergon cuspidatum, Campylium stella-
...m* and *Brac...ythecium rutabulum* occur in the community, sometimes in abundance over litter or patches of bare substrate, bryophytes are typically not well represented.

Seedlings and saplings of *Salix cinerea* and *Alnus glutinosa* are occasional to frequent throughout the community and, in some sub-communities, are characteristically abundant giving a scrubby appearance to the vegetation. Other generally low-frequency woody species include *Frangula alnus* and *Betula pubescens* and, in one type of *Peucedano-Phragmitetum*, *Myrica gale* forms a distinctive open canopy.

Sub-communities

Seven sub-communities are characterised here. The first two, though they preserve many of the general floristic features of the community, are superficially more distinctive, being diagnosed partly on the dominance of *Carex paniculata* and *Glyceria maxima*, two species which make a negligible contribution elsewhere in the *Peucedano-Phragmitetum*. The five remaining sub-communities are characterised mainly on differences in the tall-herb and sedge/rush layers and, even within individual sub-communities, can be dominated by a variety of helophytes.

***Carex paniculata* sub-community**: Tussock swamp Pallis 1911 *p.p.*; *Caricetum acutiformo-paniculatae* V1. & Van Zinderen Bakker 1942 *p.p.*; Primary tussock fen Lambert 1951; *Caricetum paniculatae thelypteridetosum* and *caricetosum acutiformis* Segal & Westhoff 1969; *Caricetum paniculatae peucedanetosum* Wheeler 1978. Here, the physiognomy is overwhelmingly dominated by the sometimes enormous tussocks of *C. paniculata*; indeed, the vegetation looks like a sedge-swamp with a *Peucedano-Phragmitetum* element perched on the tussock tops. The tussocks themselves are frequently up to 1 m or more in height and diameter and from among the spreading crown of shoots grow many of the characteristic tall dicotyledons of the community, *Lysimachia vulgaris*, *Lythrum salicaria*, *Eupatorium cannabinum*, *Peucedanum palustre* and the sprawling *Galium palustre*. There may also be some *Solanum dulcamara*, *Epilobium hirsutum*, *Scutellaria galericulata*, *Calystegia sepium* and *Lycopus europaeus* and, particularly distinctive of this sub-community, *Scrophularia auriculata* and the naturalised alien *Impatiens capensis*. *Thelypteris palustris* is also frequent, sometimes forming a dense cover of fronds. Smaller herbaceous species frequent in other sub-communities here tend to be rare but there may be a patchy mat of *Calliergon cuspidatum*, *Brachythecium rutabulum* and *Lophocolea bidentata sensu lato* over the shoot bases. This is also one of the types of the *Peucedano-Phragmitetum* that is especially prone to invasion

by trees and seedlings and saplings of *Salix cinerea* and *Alnus glutinosa* are often abundant among the herbs.

This sub-community seems to develop by *C. paniculata* gaining a hold on the felted root mat of old *Phragmitetum australis* (Lambert 1951: see below) and reed may remain quite abundant between the expanding tussocks. However, *Calamagrostis canescens* and *Cladium mariscus* and the two most important components of the *Peucedano-Phragmitetum* sedge/rush layer, *Juncus subnodulosus* and *Carex elata*, are generally uncommon. Here, these are sometimes replaced by a variety of tall swamp helophytes and sedges of medium stature which occur patchily in the wet hollows that develop as the bulky tussocks begin to depress the underlying substrate, e.g. *Typha angustifolia*, *Sparganium erectum*, *Carex pseudocyperus*, *C. acutiformis* and *C. riparia*. *Rumex hydrolapathum* and *Ranunculus lingua* may also be locally prominent as emergents. In other cases, open water has just a floating mat of *Lemna minor* or is completely devoid of vegetation.

***Glyceria maxima* sub-community:** Yare Valley fen Pallis 1911; *Glycerietum maximae* Hueck 1931 *p.p.*; Primary *Glycerietum* hover fen *p.p.* and Secondary *Glycerietum* fen Lambert 1946; Mowing marsh vegetation Lambert 1948 *p.p.*; *Peucedano-Phragmitetum glycerietosum* Wheeler 1980*a*. The somewhat varied vegetation included in this sub-community is characterised in general by the prominence of *G. maxima* among the taller fen species. In some stands, it is dominant and its lush growth and the lodging of the shoots tend to depress the richness of the associated flora. *Phragmites* persists, together with most of the tall dicotyledons of the community, and there may be patches of *Epilobium hirsutum* and *Urtica dioica*, some *Caltha palustris* and *Scutellaria galericulata* below and sprawling *Calystegia sepium*. Other stands have a reduced cover of *G. maxima* and here *Phragmites* is normally the dominant in vegetation that is substantially richer, especially in the sedge/rush layer. *Carex elata* and, especially, *Juncus subnodulosus* are more plentiful and some of the typical sedges of the *C. paniculata* sub-community occur here too, sometimes in abundance, with *C. paniculata* itself, *C. riparia* and *C. acutiformis* attaining locally dominant roles. Sprawlers are also more conspicuous with *Galium palustre*, *Solanum dulcamara*, *Vicia cracca* and the rare *Lathyrus palustris*. The last, together with *Thalictrum flavum*, are particularly distinctive of this and the next sub-community within the *Peucedano-Phragmitetum*. Smallers herbs, too, are better represented in this more open vegetation with *Mentha aquatica*, *Myosotis scorpioides* and sometimes *Impatiens capensis*. *Calliergon cuspidatum* and *Brachythecium rutabulum* occur patchily over litter and bare substrate.

***Symphytum officinale* sub-community**: Wicken Fen species groups 7–13   Yapp 1908; Wicken Fen Mixed Sedge   Godwin & Tansley 1929 *p.p.*; Woodwalton Fen *Calamagrostis* sere   Poore 1956*b*; Catcott Heath fen Willis 1967 *p.p.*; *Peucedano-Phragmitetum symphytetosum*   Wheeler 1980*a*. In general floristics, this is much more of a central type of *Peucedano-Phragmitetum* than either the *Carex paniculata* or the *Glyceria* sub-community, although it shows some particular similarities to the latter. Often, however, local variation, especially among the dominants, masks the general features of the vegetation. At Wicken Fen, for example, *Phragmites* and, especially, *Cladium* are prominent (Yapp 1908, Godwin & Tansley 1929, Wheeler 1975); at Woodwalton Fen, *C. canescens* and, more locally, *C. epigejos* are often dominant (Poore 1956*b*, Wheeler 1975). Among this canopy, though, all the tall dicotyledonous constands of the community are well represented with, in addition, frequent *Iris pseudacorus*, *Lycopus europaeus*, *Angelica sylvestris*, *Cirsium palustre*, *Thalictrum flavum*, *Stachys palustris* (at Woodwalton) and, especially distinctive here, *Symphytum officinale*.

The sedge/rush layer is sometimes extensive, though generally species-poor and, like the helophyte element, variable in its composition. There is usually some *Carex acutiformis* but *Juncus subnodulosus* can vary from being abundant (as in some Wicken stands) to very local or absent (as at Woodwalton). A distinctive feature in some cases, especially where there is little *Cladium*, is the prominence of tussocks of *Molinia caerulea*, often with some *Carex panicea* and *Salix repens* (Wicken); *M. caerulea* was also prominent, with *J. subnodulosus*, in the tall-herb fen vegetation very similar to this sub-community described from Catcott Heath on the Somerset Levels (Willis 1967). There, too, *Thelypteris palustris* was abundant, though it is generally rare in this sub-community.

Smaller herbaceous species are not well represented here, though there are usually some sprawlers with *Galium palustre*, *Lathyrus palustris* and *Calystegia sepium* frequent; at Wicken, *Galium uliginosum*, generally rare in the *Peucedano-Phragmitetum*, is common. Bryophytes are somewhat variable: *Calliergon cuspidatum* and *Campylium stellatum* occur throughout and, in wetter sites, may be prominent; drier stands often have only a few strands of *Brachythecium rutabulum*.

In general, this vegetation is drier underfoot than many types of the community and invasion by woody species is frequent. *Salix cinerea* seedlings and saplings occur commonly, sometimes with *Crataegus monogyna*, *Betula* spp. and *Frangula alnus*. These, together with the varying pattern of dominance and the patchy occurrence of bulky tall herbs, combine to give the vegetation a somewhat scruffy appearance.

Typical sub-community: Bure Valley fen   Pallis 1911; Primary *Phragmitetum p.p.* and Secondary *Phragmitetum*   Lambert 1946; Mowing marsh vegetation   Lambert 1948 *p.p.*; Anthropogenic *Phragmitetum*, *Calamagrostis* and *Juncus* fens   Lambert 1951; *Peucedano-Phragmitetum typicum*, typical variant Wheeler 1978. *Phragmites* is the most frequent helophyte here and is usually dominant. *C. canescens* is less common and variable in abundance, being particularly prominent in drier situations such as the well-drained edges of marsh compartments but completely absent from wetter stands. *Cladium* is much less frequent than either of these species, although it may attain local abundance. Tall dicotyledons are well represented with, in addition to the community constants, *Valeriana officinalis*, *Iris pseudacorus*, *Lycopus europaeus*, *Rumex hydrolapathum*, *Angelica sylvestris* and *Cirsium palustre*. *Epilobium hirsutum*, *Phalaris arundinacea* and *Urtica dioica* are uncommon and rarely abundant.

There is usually a prominent understorey of *J. subnodulosus* with varying amounts of *Thelypteris palustris*; in some stands, especially where there is a shrubby element, the latter species forms a thick cover. Sedges are, however, rather infrequent: there is sometimes a little *C. elata* or *C. pseudocyperus* and, more rarely, *C. appropinquata* or *C. acutiformis*. Beneath is a generally species-poor small-herb element: *Mentha aquatica*, *Scutellaria galericulata*, *Potentilla palustris* and *Hydrocotyle vulgaris* occur here, sometimes with *Oenanthe fistulosa*. *Galium palustre* is the only common sprawler. *Calliergon cuspidatum* and *Campylium stellatum* are frequent but rarely abundant. In drier stands, seedlings and saplings of *Salix repens* and *Alnus glutinosa* may be prominent.

***Cicuta virosa* sub-community:** *Peucedano-Phragmitetum cicutetosum*   Wheeler 1978. Again, *Phragmites* is usually the most prominent helophyte although *C. mariscus* is also frequent and it may dominate. *C. canescens* is less common here than in the Typical sub-community and rarely abundant; in some stands, it tends to be replaced by *Typha angustifolia* which is preferential for this sub-community. Among the tall dicotyledons, umbellifers are especially prominent: as well as *Peucedanum palustre*, *Sium latifolium*, *Berula erecta* and *Cicuta virosa* are also constant here and, from July to September, give a characteristic white bloom to uncut stands. Many of the other *Peucedano-Phragmitetum* tall herbs remain frequent including *Lysimachia vulgaris*, *Lythrum salicaria*, *Eupatorium cannabinum*, *Iris pseudacorus* and *Lycopus europaeus*. *Ranunculus lingua* is also preferentially constant here. Some species more characteristic of drier fens, such as *Filipendula ulmaria*, *Valeriana officinalis* and *Angelica sylvestris*, are markedly reduced in frequency.

The sedge/rush and small herb layers are typically well developed and species-rich. *J. subnodulosus* and *C. elata* are both constant and each may be abundant and, in addition, *C. pseudocyperus* is preferential here and often noticeable by virtue of its yellow-green shoots. A variety of other sedges of medium size may make a locally important contribution (see variants below). Among smaller dicotyledons, *Mentha aquatica*, *Potentilla palustris*, *Hydrocotyle vulgaris* and *Oenanthe fistulosa* are constant and *Cardamine pratensis* and *Stellaria palustris* occasional. *Galium palustre* occurs throughout. Bryophytes are somewhat variable although *Calliergon cuspidatum* and *Campylium stellatum* are constant. Bushes of *Salex cinerea* and *Myrica gale* occur occasionally.

Typical variant: *Peucedano-Phragmitetum cicutetosum*, typical variant   Wheeler 1978. In this more species-poor vegetation, *Phragmites* is usually obviously dominant, often with some *Cladium* and *C. canescens*. In wetter stands, *Carex paniculata* or *C. riparia* may attain local prominence. The sedge/rush layer is here less bulky and rich although there is generally some *J. subnodulosus*, *C. elata* and *C. pseudocyperus*. Where *Filipendula ulmaria* and *Valeriana officinalis* occur in this sub-community, it is in this drier variant and here, too, sprawlers may be more conspicuous with *Solanum dulcamara* and *Calystegia sepium*.

Carex lasiocarpa variant: *Cicuto-Caricetum pseudo-cyperi comaretosum*   Van Donselaar 1961 *p.p.*; *Peucedano-Phragmitetum cicutetosum, Carex lasiocarpa* variant   Wheeler 1978. Here, *Cladium* and *Calamagrostis canescens* are less frequent and abundant and, although *Phragmites* may remain prominent and be accompanied by *Typha angustifolia*, it is the sedge/rush layer that often comprises the bulk of the vegetation. *J. subnodulosus*, *C. elata* and *C. pseudocyperus* are here accompanied by *C. lasiocarpa*, which may form a thick understorey, and more occasionally by *C. appropinquata* and *C. diandra*. *C. rostrata* has its rare Broadland station in vegetation of this kind and *C. limosa*, although not refound by Wheeler (1978), also appears to have occurred in Norfolk in this variant. Beneath, smaller herbs are often numerous and abundant: in addition to those characteristic of the sub-community in general, there are frequent clumps of *Menyanthes trifoliata* and *Caltha palustris* and scattered shoots of *Equisetum fluviatile* and *Pedicularis palustris*. The orchids *Epipactis palustris*, *Dactylorhiza incarnata* and *D. majalis* ssp. *praetermissa* occur occasionally and *Alisma plantago-aquatica*, *Baldellia ranunculoides* and *Nymphaea alba* are sometimes found on the oozing peat. Much of the ground surface may, however, be clothed with a carpet of bryophytes studded with the shoots of *Hydrocotyle vulgaris*. As well as *Calliergon cuspidatum* and *Campylium stellatum*,

there are records for *Calliergon giganteum*, *Riccardia pinguis* and, less frequently, *Bryum pseudotriquetrum*, *Scorpidium scorpioides*, *Plagiomnium elatum* and *Campylium elodes*.

Schoenus nigricans sub-community: *Peucedano-Phragmitetum schoenetosum*   Wheeler 1978. Although *Cladium* and, less frequently, *Phragmites*, normally dominate in this sub-community, the helophyte and tall-herb components are somewhat poorer than in other types of *Peucedano-Phragmitetum*: *Calamagrostis canescens*, *Filipendula ulmaria*, *Angelica sylvestris*, *Cirsium palustre* and *Rumex hydrolapthum* are all, at most, occasional. Beneath, however, the sedge/rush layer is distinctive in the high frequency, and sometimes the abundance, of *Schoenus nigricans*, intermixed with *C. elata* and *J. subnodulosus*. The small herb element, too, has a characteristic quality, although it is somewhat variable: as well as the more usual *Mentha aquatica* and *Hydrocotyle vulgaris* (which are constant here) and *Scutellaria galericulata* and *Oenanthe fistulosa* (which occur occasionally), there is often some *Ranunculus flammula* and *Carex panicea*, sometimes with *Molinia caerulea* and *Salix repens*. Occasional richer areas, usually of small extent and with a lower growth, may have scattered plants of *Pedicularis palustris*, *Carex lepidocarpa* and *Epipactis palustris*. *Osmunda regalis* has been recorded in this sub-community at one site. Along the Thurne valley in Norfolk, this kind of vegetation is additionally distinctive in the presence of numerous plants of *Oenanthe lachenalii* and occasional *Samolus valerandi*. Here, too, there are often abundant lianes of *Calystegia sepium* and sometimes a low carpet of *Agrostis stolonifera*. Bryophytes are typically not well developed.

Myrica gale sub-community: *Cladium* anthropogenic fen   Lambert 1951; *Peucedano-Phragmitetum myricetosum*   Wheeler 1978. In the rather species-poor vegetation of this sub-community, *Phragmites*, *Cladium* or *Calamagrostis canescens* normally dominate. The constant dicotyledons of the community are all well represented but, among other tall herbs, only *Valeriana officinalis* is at all frequent. The sedge/rush layer, too, is not rich: *J. subnodulosus* is generally present, though often in small amounts, and *C. elata* and other sedges are rare or absent. *Thelypteris palustris*, however, is often prominent and there is sometimes a little *Molinia caerulea*. The most distinctive feature of the vegetation, though, is the presence of scattered leggy bushes of *Myrica gale*, often with saplings of *Betula pubescens*, *Salix cinerea* and *Alnus glutinosa*. Beneath this uneven shrubby canopy, smaller herbs and bryophytes are typically sparse: even *Mentha aquatica* is no more than occasional, although small amounts of *Salix repens* are characteristic.

Habitat

The *Peucedano-Phragmitetum* is generally restricted to fen peats with a moderate to high summer water-table and some winter flooding with base-rich, calcareous and often oligotrophic waters (Wheeler 1978, 1980*a*, 1983). It is a community of topogenous mires, occurring as primary fen in open-water transitions but being especially characteristic of flood-plain mires where long and complex histories of exploitation for peat and marsh crops have often led to the maintenance of the community as anthropogenic secondary vegetation.

The community is now almost entirely confined to Broadland with a few outlying stations on fragments on flood-plain mires that have escaped drainage and embankment or where a high level of unenriched waters is artificially maintained. Although it was certainly more extensive in the past (e.g. Pallis 1911), it is probably essentially a vegetation type of those areas to the south and east where once extensive flood-plains with calcareous catchments occur in a more continental climate. Some of the most important species in the community, such as *Cladium* and *Juncus subnodulosus*, have a well-established preference for base-rich, waterlogged conditions and others have a Continental or Continental Northern distribution, being generally limited in Britain to wet (though not necessarily base-rich and calcareous) habitats away from the west of the country: e.g. *Ranunculus lingua*, *Rumex hydrolapathum*, *Carex appropinquata*, *C. diandra*, *C. lasiocarpa*, *Cicuta virosa*, *Lathyrus palustris*, *Peucedanum palustre*, *Stellaria palustris*, *Thelypteris palustris* (Matthews 1955, Ratcliffe 1977, Jermy *et al.* 1978).

Within the *Peucedano-Phragmitetum*, many of the floristic differences between the sub-communities seem to be related most clearly to edaphic conditions, particularly the water regime and the trophic state of the substrate. Where the community occurs as primary fen, mostly in the flooded medieval peat diggings of the Broads, these variables are largely dependent upon the progress of the natural hydrarch succession around the open waters but, elsewhere, shallow peat extraction and other forms of soil disturbance have much altered the state of the surface, often producing very distinct local conditions for recolonisation. On the more solid peats, too, the community has been and, on a much reduced scale, still is affected by a variety of mowing treatments. The most obvious impact of these seems to be on the quantitative balance between species within the sub-communities affected, thus producing an overlay of variation in dominance upon the more complex, edaphically influenced differences (Wheeler 1978, 1980*a*) but consistent treatments may well contribute to qualitative

variation between sub-communities (e.g. Godwin 1941; see below).

There are other reasons, too, why it is very difficult to disentangle the influences of soil and treatment variables on differences in composition and structure between and within the various sub-communities. If there is some measure of primary edaphic control on the floristics of the community, then treatment options may well be limited and particular mowing regimes adopted initially to suit the species growing best naturally (Wheeler 1978). Local traditions of treatment may thus have tended to confirm local natural effects and it may be such combinations of influences that are responsible for the rather striking pattern of distribution of some of the sub-communities in particular stretches of fen. Mowing, at least winter mowing for reed, has also traditionally had to be confined to drier areas of fen which means that, in some cases, such interactions between soil and treatment variables have occurred over only part of the range of water and nutrient regimes experienced by the community. Where mowing has occurred, it is likely, too, that the treatment itself has produced some modification of soil conditions with harvesting of the crop, burning of litter, controlled unseasonal flooding and the application of fertilisers or marl. Making sense of all this is not helped by the sometimes fragmentary survival of the community, even in Broadland, and the difficulty of documenting treatment histories, especially now that most of the mowing traditions are defunct. The following therefore provides little more than an indication of the possible habitat relations of the community.

Typically, the *Peucedano-Phragmitetum* occurs on raw peat soils with, in Broadland, a water pH generally between 6.5 and 7.5 and dissolved calcium levels in the range 60–120 mg l^{-1} (Wheeler 1983). Where the community develops as primary fen around the Broads margins, the peat is the often still fairly thin layer of organic material that has been accumulating over the rhizome and root mat of the preceding swamp vegetation and the fen is subject to edaphic conditions which continue to be determined largely by the nearby open waters. The peat is often continuously waterlogged throughout the year and closely exposed to any fluctuations in water-level and to irrigation with any dissolved salts. In such situations, too, the vegetation is inaccessible to grazing stock (though not necessarily to coypu and wildfowl: see below) and often impossible to mow, especially in winter. In general, therefore, the floristics and structure of such primary *Peucedano-Phragmitetum* tend to be strongly influenced by the continuing natural persistence of the one-time swamp dominant. Which particular helophyte this it seems to depend partly on the nutrient status of the waters.

Along the Bure and the Ant valleys, it is the Typical

sub-community which comprises the bulk of primary fen with *Phragmites* as the usual dominant over a firm, though sometimes floating, peat raft with a summer water-table around the substrate surface and with often prolonged winter flooding. Much less commonly, *C. mariscus* dominates in essentially similar vegetation: this species is more prominent around isolated, land-locked broads like Upton where it perhaps has the edge on *Phragmites* in more oligotrophic conditions (Lambert 1951, Phillips 1977, Wheeler 1980a). The tall dicotyledons so typical of the *Peucedano-Phragmitetum* may also find it more difficult to gain a hold in *Cladietum* swamp with its very dense shoot cover and often thick litter layer, though the effects of these may be offset somewhat if *Cladietum* is mown: cutting for sedge, being a summer operation carried out when the surface is drier, has perhaps pushed further towards open water than mowing for other crops.

In more eutrophic conditions, *Glyceria maxima* can retain its pre-eminence over *P. australis* in primary fen as well as in swamp (Lambert 1946) and, along the Yare, with its pronounced pattern of tidally influenced irrigation, it is the usual dominant in the fens around the broads and dykes in the *Glyceria* sub-community. This kind of *Peucedano-Phragmitetum* is floristically distinct in the presence of other more nutrient-demanding species, such as *Epilobium hirsutum* and *Typha latifolia*, and it also has frequent *Carex riparia* which is perhaps able to tolerate the often loose substrate: typically, primary fen of the *Glyceria* sub-community develops on a quaking raft of *G. maxima* rhizomes, peat and deposited silt which rises and falls with the fluctuating waters (the 'hover fen' of Lambert 1946). *G. maxima*, however, retains its overwhelming dominance here only so long as there is an adequate supply of nutrients from the freely circulating waters or from more mineral-rich substrates. Lambert (1946, 1948; see also Buttery & Lambert 1965) showed how, around the broads and dykes of the middle Yare, the abundance of the species declined in moving to the centres of fen compartments on peat. With the increased dereliction of dykes and consequent restriction of free water movement since the time of Lambert's surveys, this *G. maxima*-dominated primary fen has become even more restricted in its distribution (Wheeler 1975). Such areas of the *Glyceria* sub-community has have not been invaded by shrubs and trees are now much more frequently dominated by *P. australis* and have a richer and often confused associated flora which reflects both the drying and impoverishment of the peat and the complex history of treatment and neglect in this area (see further below).

The *Carex paniculata* sub-community shares certain floristic features with the *Glyceria* sub-community and it, too, may be characteristic of more eutrophic substrates (Lambert 1951). The edaphic conditions are,

however, rather distinctive because, as the *C. paniculata* tussocks enlarge upon the usually firm but often swinging mat of *P. australis* peat where they have gained a hold, they depress it, recreating swampy hollows between them (Lambert 1951, Wheeler 1978, 1980a). It is on the drier sides and tops of the tussocks themselves that much of the *Peucedano-Phragmitetum* flora establishes itself and here the vegetation is very prone to the invasion of woody species which may quickly form a canopy and curtail a primary fen phase.

The *Cicuta* sub-community also occurs as primary fen around some broad margins in the Ant valley, although in this kind of situation it is usually fragmentary and built around scattered *C. paniculata* tussocks (Wheeler 1980a). Elsewhere, however, the *Carex lasiocarpa* variant of this sub-community can be found as a striking kind of unmown fen vegetation in generally small stands in very wet hollows. Sometimes this vegetation occurs as a swinging raft of 50–100 cm of interlacing rhizomes over a suspension of semi-fluid peat and silt, suggesting that it has developed by the closing over of shallow depressions such as those left by nineteenth-century peat digging as on Catfield Fen (Wheeler 1975, 1978). The prominence in this variant of sedges which are also characteristic of the *Carex rostrata-calliergon* fen suggests that a combination of extreme wetness with lack of throughput may play some part in determining the distinctive nature of the associates but the significance of these conditions is unknown (Wheeler 1980c).

Away from the vicinity of open water, the peat is generally drier with a lower summer water-table and less prolonged winter flooding, if any, and is further removed from any natural enrichment by eutrophic river waters (Giller 1982, Wheeler 1983). In such situations, the vigour of some of the helophytes prominent in certain types of primary *Peucedano-Phragmitetum*, e.g. *Glyceria maxima*, *Typha* spp., may decline and other species more tolerant of drier conditions, e.g. *Juncus subnodulosus*, *Calamagrostis canescens*, become more abundant. Here too, however, in contrast to its often treacherous condition nearer the open water, the community has been accessible for the harvesting of various fen crops and there is little doubt that much *Peucedano-Phragmitetum* has been maintained as secondary fen by a range of cutting regimes which have repeatedly set back the invasion of woody species possible with a lower water-table. Of those forms of the community which at present extend into open-water transitions as primary fen, the Typical and *Glyceria* sub-communities seem able to survive such treatment on drier peats with some modification to their floristics and physiognomy but with their basic integrity intact, except in the most extreme conditions. The *Carex paniculata* sub-community seems to have been rarely mown in its fully-developed natural form because of the obstruction caused by the bulky sedge-tussocks but it is known that these were sometimes dug out to facilitate cutting and, in Broadland, they found a quaint local use, when trimmed down, as fireside seats and church kneelers (Lambert 1965).

Two further types of *Peucedano-Phragmitetum*, the Typical variant of the *Cicuta* sub-community and the *Schoenus* sub-community, are generally found only on these drier peats and then only where there has been a tradition of mowing. It is consequently particularly difficult here to say what balance of edaphic and treatment variables has influenced their composition. The former occurs on fairly firm peats around Barton Broad and on Woodbastwick Fen where the summer water-table is at least 20 cm below the surface (Wheeler 1975). It has perhaps developed in these sites over shallow nineteenth-century peat diggings which have now acquired a solid infill of new organic matter consolidated by repeated cropping of *Phragmites* (Lambert 1951). The *Schoenus* sub-community is extensive in the Ant and Thurne valleys in similarly drier situations but, where there has been peat digging, it occurs only on the uncut surfaces (Wheeler 1980a). The distinctive stands of this sub-community around the Thurne broads also show some maritime influence with *Oenanthe lachenalii* and *Samolus valerandi* occurring, perhaps in response to percolation of saline water under the fairly narrow bank of dunes that separates these fens from the sea (Wheeler 1980a).

Mowing has also played some part in influencing the vegetation of the *Symphytum* sub-community but also of importance here is the marked association with the driest and sometimes most disturbed peat surfaces on which the *Peucedano-Phragmitetum* occurs (Wheeler 1975, 1980a). This sub-community is found on flood-plains outside Broadland where, now set in drained and improved agricultural landscapes, small fragments of mire remain, their drying peat surfaces often marked by complex patterns of disturbance including peat digging, ditching, the grubbing up of scrub and carr, marling, eutrophication by fertilisers or from farmland run-off and ploughing. Here the *Symphytum* sub-community survives as secondary vegetation, now maintained by renewed mowing, but exhibiting some, often strikingly site-specific, loss of rich-fen species and some gain in plants of disturbed and enriched situations which bear testimony to the long history of interference. This is most noticeable at Woodwalton Fen where the sub-community occurs over the ridge-and-furrow of old cuts in fen peat which seems to have been exposed by stripping of overlying acid peat, then inundated by nutrient-laden water from surrounding farmland and then to have dried (Poore 1956b). The rich-fen element in its flora is markedly impoverished (although *Peucedanum palustre*, together with *Sonchus palustris*, was intro-

duced in the 1920s: Poore 1956b, Duffey 1971) and aggressive species, such as *Calamagrostis epigejos* which is not a fen plant at all (Poore 1956b), have appeared as chance colonists and gained a vigorous hold. At Wicken Fen, where there has been less drastic disturbance and a possibly more consistent tradition of mowing, the *Symphytum* sub-community appears more like other forms of secondary *Peucedano-Phragmitetum*. It here includes some of the vegetation which develops during the mowing cycle on peats with a summer water-table at least 30 cm below the surface (Godwin 1931) and where there has probably been some sporadic peat digging and removal of carr (Godwin 1941, 1978; National Trust 1947; Wheeler 1975). Fragments of fen vegetation akin to the *Symphytum* sub-community have also been recorded from Catcott Heath on the Somerset Levels (Willis 1967) and from Kent (Rose 1950), again isolated within much-altered landscapes and perhaps with similarly peculiar site histories.

It is easy to make general suppositions about the effects of mowing on these sub-communities of the *Peucedano-Phragmitetum* but much more difficult to be precise. Even when mowing was a normal part of the Broadland rural economy (and at one time or another most of the area at present occupied by the community was probably regularly cut: Lambert 1946, 1951, 1965), stretches of fen seem to have been subjected to a complex pattern of treatments which might vary from year to year, from one compartment to the next and even within compartments according to need, available labour, weather conditions and changes of ownership or commonland use. Even at the time of Lambert's (1946) survey of the Yare fens, when there was still some memory of active fen exploitation, it was generally difficult to trace treatment history back for more than 10–20 years. Now, with the almost universal neglect of fens as a source of marsh crops, such local knowledge is largely forgotten and many areas once mown have passed to scrub and carr. In other sites which have long attracted particular interest, such as Woodwalton Fen, treatment records are imperfect (Poore 1956b) and the pioneering work of Godwin (1941), in the rather atypical conditions at Wicken Fen, remains the only published experimental consideration of the effects of mowing on this kind of fen vegetation.

The *Peucedano-Phragmitetum* has traditionally provided four types of crop: *Phragmites* for reed thatch, *Cladium* for sedge thatch, *G. maxima* for green fodder, winter hay or litter and more mixed vegetation, often with abundant *Juncus subnodulosus* and/or *Molinia caerulea*, also for litter. The most obvious general effect of regularly mowing for any of these crops was to set back repeatedly the natural invasion of woody species and so maintain the community, albeit in modified forms as secondary fen. This would probably have been aided by the keeping of dykes in good repair for water circulation and easy access and in some cases also by controlled flooding, both of which could maintain the substrate in a moister condition than might otherwise have been the case.

Beyond this, mowing seems to have introduced into the community some measure of interchangeability of dominance, still dependent to some extent on edaphic conditions, but controlled to a considerable degree by the timing of the cut. Different mowing regimes are appropriate for each of the crops and, although choosing to mow for one or the other may have capitalised on a natural abundance developed in response to edaphic conditions, cutting at a particular time could select for the continuing abundance of the desired crop and the demise of possible competitors. This is best seen in the case of *Phragmites* and *Cladium*, the two most important traditional crops and the most widespread dominants of the community. Although both these species, and especially the former, have been available for cutting in their often almost pure swamp communities, there is no doubt that they have been selected for in the *Peucedano-Phragmitetum*, particular stands having been treated as either reed-beds or sedge-beds. Mowing for reed has traditionally been carried out in winter when the annual crop of stems so ideal for thatching is dead and it has normally been repeated annually or biennially with no deleterious effect (Haslam 1972a, b; McDougall 1972). *Cladium*, a slow-growing evergreen, is especially susceptible to such treatment and will quickly decline if it is persisted with (Godwin 1929; Conway 1936a, 1937a, 1942). The best period for harvesting sedge is early summer (T. Rowell, personal communication) and, although crops have traditionally been taken only every three to five years to allow adequate recovery (Godwin & Tansley 1929, Godwin 1941, McDougall 1972), this is the time of year when *Phragmites* is most vulnerable to the removal of its flush of green shoots (Haslam 1972a). Within some of the sub-communities of the *Peucedano-Phragmitetum* (notably the Typical and *Symphytum* sub-communities), it is common to find stands which differ only in the proportions of *Phragmites* and *Cladium* which suggests that, although both can naturally occur, treatment has played some part in shifting the balance towards one or the other. Even in those sub-communities where there is a more consistent pattern of dominance (e.g. *Phragmites* in the Typical variant of the *Cicuta* sub-community and *Cladium* in the *Schoenus* sub-community), it is not possible to rule out treatment effects entirely.

Along the Yare, treatment has probably also played some part in determining the composition of the *Glyceria* sub-community. Here, *G. maxima* is much more obviously a natural dominant in both swamp and primary fen provided eutrophic conditions are maintained, although it has generally been inaccessible for mowing in the former where it occurs as swinging and often semi-submerged rafts. From primary fen, *G. maxima* was traditionally summer-cut, usually before panicle emergence in June for green feed or winter hay and sometimes again, as part of a litter crop, between July and October (Lambert 1946). Indeed, there is some evidence that as many as three mowings could be made in a single season where growth of the continuously produced summer shoots was especially prolific (Lambert 1946, 1947b). Again, such treatment was inimical to the vigour of any *Phragmites* that gained a natural hold in the less eutrophic areas where it might have an edge on *G. maxima* and could perhaps eliminate it altogether (Lambert 1946). Certainly, with the virtually universal relaxation of cutting in the fens along the Yare, the spread of *Phragmites* as a dominant in the *Glyceria* sub-community is very pronounced, although this is probably also partly due to the neglect of dykes and the consequent restriction of water circulation.

In the days when mowing was in full swing, such treatment differences as these contributed to the often bewildering variety of dominance in particular stretches of fen. Now, much of this variegation is masked by a more uniform advance of scrub and carr over former mowing marsh but evidence from former days and such stands as have a more recent cutting history suggest that these mowing regimes have not been ultimately responsible for the overall floristic differences which characterise the sub-communities. This is not to say that these kinds of regimes have had no effects other than to deflect succession or to alter simple patterns of dominance but it seems that, in general, these impacts have not disrupted the edaphically influenced variation within the community (Wheeler, 1978, 1980a). One effect that does seem to be discernible is an increase in species-richness associated with the regular but infrequent summer mowing for sedge. Indeed, this may contribute to the general species-richness of the community as a whole since sedge moving has been quite widespread but it is also visible in a more particular way within the community. Winter mowing for reed, although perhaps reducing the cover of other evergreen species apart from *Cladium*, e.g. *Carex paniculata*, *C. riparia*, *C. acutiformis*, might be expected to have little impact on the majority of associates which die down completely in winter and which could perhaps be capable of vigorous growth on the open litter-stripped surface of the mown bed in the following spring (Lambert 1951). Such mowing can, however, produce a marked increase in the density of *Phragmites* by exposing it to the effects of frost (Haslam 1972a, b) and light attenuation in reed-dominated fen vegetation like the *Peucedano-Phragmitetum* can be severe (van der Toorn 1972, Wheeler & Giller 1982a). Both mown and unmown stands of the community dominated by *Phragmites* can thus be similarly rather species-poor.

Stands which have been mown every four years or so for sedge, on the other hand, are noticeably more species-rich than either unmown sedge-beds (which tend towards the *Cladietum marisci*: see below) or reed-dominated stands of the *Peucedano-Phragmitetum*, not only in vascular associates but also in the variety of bryophytes (Wheeler & Giller 1982*a*). This may be because, with the removal of the bulky sedge growth and the accumulated litter (both have traditionally been harvested) and the subsequent slow regrowth of the *Cladium*, there is much more opportunity for the estab-lishment of a varied understorey (Godwin 1941, Wheeler & Giller 1982). Moreover, some of the species which are often prominent in mown sedge-beds, e.g. *Juncus subnodulosus, Schoenus nigricans, Molinia caeru-lea*, are tolerant of summer mowing and readily respond to it by producing new shoots (Godwin 1941, Lambert 1951, 1965).

Summer mowing of *G. maxima*, by contrast, does not seem to have favoured an increase in the diversity of the *Glyceria* sub-community in wetter, more eutrophic sit-uations along the Yare (Lambert 1946). Even though this was often undertaken annually, the ready produc-tion of new shoots and their lodging seems to have militated against much richness in the associated flora. Away from the influence of enriched waters, however, where the vigour of *G. maxima* declined, summer mow-ing was often accompanied by an increase in other species, notably *J. subnodulosus* (Lambert 1946).

The ability of this species and *Molinia caerulea* to flourish under a regime of annual summer mowing, where other possible competitors were excluded either by the timing and frequency of the cut or by less eutrophic conditions, often enabled a yearly crop of litter vegetation to be taken from stands of the *Peuce-dano-Phragmitetum* in drier situations (Lambert 1946, 1951, 1965). This kind of treatment can certainly produce gross changes in the community over a fairly short period, converting it to a wholly different vege-tation type (Godwin 1941).

The most obvious anthropogenic effect in the community now is due to neglect with the discontin-uance of mowing and the natural silting and over-growth of unmaintained dykes. It is difficult to separate the influences of these (Lambert 1946) but the reversion of fen to woodland is clearly visible throughout Broad-land and is reflected in the floristics of the community in two ways. First, there is the general abundance of saplings, especially of *Salix cinerea* and *Alnus glutinosa*, throughout most of the sub-communities. It is possible, too, that the prominence of other species is related to neglect: *Thelypteris palustris*, for example, seems to be more characteristic of stands with frequent bushes, and

Calamagrostis canescens, which responds to the cess-ation of mowing by producing a pronounced tussock habit (Lambert 1951) is nowadays often very abundant. Second, there is the particular association of the *Myrica* sub-community with derelict mowing marsh, especially in the Ant valley and around Upton and Woodbastwick Fen, where *M. gale* can form a vigorous growth of bushy canopy, often accompanied by low *Salix repens* and frequently saplings of *Betula pubescens*, on peats with a summer water-table 10–40 cm below ground (Wheeler 1975, 1978, 1980*a*).

Two other fairly recent developments are also having some influence on the floristics of the *Peucedano-Phrag-mitetum*. There is no evidence to suggest that the well-attested eutrophication of some of the waters of Broad-land (e.g. Mason & Bryant 1975) has had any profound effect on wide areas of the community (Wheeler 1978) but, in some sites, for example around the Ant broads, there has certainly been a spread of species-poor tall-herb vegetation among the fens and *Epilobium hirsutum* in particular is now a prominent feature of some stands of the *Carex paniculata* sub-community, so luxuriant in places that some tussocks appear to be dying (B.D. Wheeler, personal communication).

The impact of coypu in Broadland has been most obvious in the swamps, in the recent marked decline of which they have probably played a major role (Boorman & Fuller 1981). The virtually complete removal of the swamp front in some areas might be expected to have a long-term effect on any natural advance of fen vege-tation which develops on the swamp raft and it has also exposed the fen more directly to erosion along its existing outer edge. There have also been some direct effects on the community itself. Certain species, such as *Cicuta virosa* and *Rumex hydrolapathum*, have shown a marked decline in some areas since the 1950s; *Lythrum salicaria*, on the other hand, is very coypu-resistant and has spread dramatically over the exposed substrate of some lows cleared of vegetation. In drier places, sus-tained grazing has produced an increase in *Agrostis stolonifera, Festuca rubra* and *Poa trivialis* or set back the invasion of *Salix cinerea*, the bushes of which are browsed and barked or loosened by having their roots gnawed (Ellis 1965).

The complex of open-water transition and flood-plain mires in Broadland, where the *Peucedano-Phragmite-tum* forms the major fen element, has a rich invertebrate fauna which has long attracted attention. Although there is no absolute coincidence between the community and organised animal assemblages, it has provided food plants or distinctive physiognomic niches for a wide variety of species, some of which are now virtually restricted to its own range, others of which have become extinct with the advance of fen reclamation. The indige-nous race of the swallowtail (*Papilio machaon britanni-*

cus) is a butterfly once fairly common throughout southern England (Ratcliffe 1977) but now confined to Broadland where *Peucedanum palustre* is the major host plant, especially for the first and larger brood of caterpillars. It has been suggested (Ellis 1965) that this species increased as *P. palustre* flourished undisturbed with the demise of mowing and burning but the recent advance of scrub and woodland into the community has made many areas less suitable for breeding and perhaps rendered the swallowtail more susceptible to natural disasters such as unfavourable weather. At Wicken Fen, the extinction of the breeding colony in 1951 has been associated with the decreasing abundance of *P. palustre* (Godwin 1978).

The British race of the large copper butterfly (*Lycaena dispar dispar*) became extinct in 1851 as its home grounds were drained or intensively mown and it became increasingly attractive to collectors. It seems to need a supply of the food plant *Rumex hydrolapathum* out of reach of flood such as is provided by some stands of the *Peucedano-Phragmitetum*. Attempts to introduce other races, *L. d. rutilus* at Wicken where *R. hydrolapathum* grows mainly along the dykes and *L. d. batavus* in the Yare valley where the fens are liable to tidal flood, have failed (Ellis 1965). At Woodwalton Fen, however, the creation of artificial open-water transitions in small saucer-shaped depressions has allowed seeded *R. hydrolapathum* to gain a hold in the optimum conditions for its germination on bare wet peat (Duffey 1968) and to grow up with other vegetation to form secondary flood-free fen which now supports a flourishing colony of *L. d. batavus* (Duffey 1971).

The community also contains some of the food plants of various wainscot moths (Ellis 1965) and stands are often rich in spiders, although the distribution of these is more a reflection of the various physiognomic niches of the vegetation than of its floristics (Duffey 1965). Some species like *Pirata piscatorius* hunt over vegetation at the water surface; others such as *Araneus cornutus* build webs among the tall flowering shoots. *Carex paniculata* tussocks have a particularly diverse and highly organised spider fauna and stacks of cut reed and especially sedge also have their distinctive populations (Duffey 1965).

Zonation and succession

In Broadland, the *Peucedano-Phragmitetum* generally occupies the middle zone of the flood-plain mires between the open water of the rivers and broads and the valley sides. In some places, it forms part of a complete and fairly clear sequence from swamp through fen to scrub and woodland, including vegetation developed over both the colonised margins of the deep peat cuts of the broads and the intact alluvial deposits behind. Such zonations have, however, been much confused,

especially on these drier solid substrates towards their landward limit, by the various traditional mowing treatments and shallow peat digging, and are now further complicated by neglect and the spread of woody vegetation over the abandoned fen compartments and dykes. Towards the open water, the transitions have been truncated somewhat by coypu activity and affected by eutrophication. The whole area occupied by these communities has, moreover, been reduced and fragmented by the progressive reclamation of the flood-plain mires for intensive farming. The *Peucedano-Phragmitetum* therefore survives here as part of sometimes extensive but often isolated tracts of complex vegetation over a largely decayed agricultural landscape now subject to the renewed but modified influence of a flood-plain mire environment.

Those few areas of the community which exist outside Broadland occur in smaller and even more isolated and modified fragments of vegetation, far removed from the natural seasonal fluctuations of large bodies of open water and generally sharply marked off from the surrounding land which has been intensively drained, widely cut over for peat and which now carries improved grassland or arable on its shrunken and wasted surface. Here, extensive zonations are rare, swamp vegetation being largely confined to fragmentary strips in the remaining open dykes and the compartments having complex patterns of fen and woody vegetation much influenced by treatment and neglect.

The natural disposition of the community between swamp on the one hand and scrub and woodland on the other is most clearly related to the water-table and, in some places in Broadland, it is possible to proceed sequentially through the various vegetation types in moving from broad margin to valley side over gently sloping ground that, in summer, is dry to increasingly greater depths and, in winter, is subject to increasingly shorter periods of inundation. Some of the classic profiles provided by Lambert & Jennings (1951) from the Bure valley show this well.

The *Peucedano-Phragmitetum* generally occupies the zone where the substrate is sufficiently dry for the overwhelming dominance of swamp helophytes to be challenged by its characteristic suite of fen species but yet sufficiently wet to prevent the permanent establishment of shrubs and trees. The extent of this zone and the sharpness of its boundaries with swamp and woodland vary according to the pattern of interaction between the fluctuating water and the topography of the flood-plain deposits, but they are also influenced by the different tolerances which species have to isolation from the open waters and their irrigating effect. Gradual switches in dominance often blur the junction between swamp and fen and also produce some natural internal heterogeneity with the sub-communities of the *Peucedano-*

Phragmitetum, as can be seen in the relationships between *G. maxima* and *Phragmites* in the *Glyceria* sub-community and *Phragmites* and *Cladium* in the Typical sub-community, and the generally increased prominence of *Juncus subnodulosus* and *Calamagrostis canescens* further away from open water (see Habitat above). The upper boundary between the community and scrub and woodland is also made rather hazy by the varieties of shade tolerance which *Peucedano-Phragmitetum* species have. Some of the important components of the community, e.g. *Carex paniculata*, *C. acutiformis*, *C. vesicaria*, *Calamagrostis canescens* and *Thelypteris palustris*, can persist under a shrub or tree cover producing a diffuse zone of overlap between fen and field layer that does not correspond exactly with the limit of the canopy.

The particular sub-communities of the *Peucedano-Phragmitetum* involved in this basic zonation seem to depend partly on the trophic state of the waters and substrate which may itself be a function of the distance from freely circulating open waters. Along the Yare, with its marked tidally influenced water movement, the usual pattern is for the *Glyceria* sub-community to occur behind a front of the *Glycerietum maximae* swamp, although, with the neglect of the open dyke system along this river, the general prominence of *G. maxima* itself and the gradual nature of the transitions between extensive areas of swamp and primary fen so characteristic of Lambert's (1946) descriptions are much less obvious. In less eutrophic conditions without any pronounced diurnal fluctuation in water-level, the community can be fronted by bands of the *Scirpetum lacustris*, the *Typhetum angustifoliae* and the *Phragmitetum australis* in progressively shallower water, although this sequence has been much fretted by coypu activity (compare the descriptions of Pallis 1911, Lambert 1951 and Boorman & Fuller 1981). In some places, there is then a zone of the *Caricetum paniculatae* which passes to the *Carex paniculata* sub-community of the *Peucedano-Phragmitetum* or, in a few localities, to fragments of the *Cicuta* sub-community. More often, however, the *Phragmitetum* gives way directly to the Typical sub-community. In the most oligotrophic conditions, the *Cladietum marisci* may be the sole swamp community fronting Typical *Peucedano-Phragmitetum* (Lambert & Jennings 1951, Lambert 1951, Wheeler 1980c). In the most natural of these kinds of sequences, the community gives way, with varying degrees of abruptness to woodland with *Alnus glutinosa*, *Salix cinerea* and *Betula pubescens* (Figure 15).

Much of the stratigraphical and observational evidence which has been used to interpret such zonations as the spatial expression of a primary hydrosere has originated from Broadland (e.g. Lambert 1946, 1948, 1951; Lambert & Jennings 1951; Walker 1970; Wheeler

1980c). Here, the *Peucedano-Phragmitetum* includes all the primary fen which develops with the gradual autochthonous accumulation of organic matter over the lake muds around the broad margins, first above the lowest limit of permanent standing water, then up towards the highest bounds ever reached on the flood, with a progressive lowering of the water-table and a reduction in the deposition of any allochthonous sediments. The present distribution of the various kinds of primary *Peucedano-Phragmitetum* in relation to the nutrient conditions of the environment suggests that these may in some way control the particular direction which the succession takes through the community (Lambert 1946, 1951; Wheeler 1980a, c, 1983; Wheeler & Giller 1982a). Lambert (1946, 1951) characterised four major pathways of succession and later observations tend to bear out their early stages except that, at the present time, *Carex acutiformis* seems to have a less distinctive role than she suggested. This species is not a very good diagnostic taxon for any sub-community and, in Lambert's own studies, it was shown to be strongly influenced by mowing (Lambert 1946, 1948) and also to be more characteristic in some places of subsequent woodland than of preceeding fen (Lambert 1951).

In great measure, the direction of the succession seems to be already set by the swamp stage. Although the distinctive fen associates of the *Peucedano-Phragmitetum* show some sorting according to the nutrient status of the waters and substrate, the swamp helophytes continue to exert some influence on the vegetation, especially on its physiognomy, and they contribute to some extent, to the particular floristic character of those sub-communities which occur as primary fen. Their growth form may also influence the speed with which the succession passes, not only from the swamps to the *Peucedano-Phragmitetum*, but also from fen to scrub and woodland. Although the absolute limit of shrub and tree invasion is probably controlled by the water-table, and especially perhaps by its maximum winter level (e.g. Godwin & Bharucha 1932, Godwin 1936), this will itself be influenced by the rate of accumulation of litter and the extent to which seedlings can gain a hold will also be affected by the density of both living and dead material on the ground. Along the Yare, for example, Lambert (1946) noted that scrub development did not generally begin in primary fen of the *Glyceria* sub-community until after the invasion of *Phragmites* when the blanketing effect of the luxuriant and lodged shoots of *G. maxima* was broken up; then abundant *Salix cinerea* appeared, followed by *Alnus glutinosa*. A dense cover of *Cladium* in Typical *Peucedano-Phragmitetum* may be similarly inimical to seedling establishment (Lambert 1951), although at Wicken it has been shown that some woody species can invade virtually pure *Cladietum marisci* directly (Godwin & Tansley 1929, Godwin &

Bharucha 1932, Godwin 1943*b*). The *Carex paniculata* sub-community, on the other hand, is especially prone to early invasion, with abundant seedlings of both *S. cinerea* and *A. glutinosa* quickly growing up on the tussock tops and forming a canopy (Lambert 1951).

The dominants of primary *Peucedano-Phragmitetum* also exert some influence on the floristics and physiognomy of the eventual woodland cover. The distinctions which Lambert (1951, 1965) made between 'swamp carr' and 'semi-swamp carr' seem to be largely structural and most of the *Alnus*-dominated woodland in which *Carex paniculata* and/or *C. acutiformis* are prominent can be incorporated within the single community of *Alnus glutinosa-Carex paniculata* woodland. The *Alnus glutinosa-Filipendula ulmaria* sub-community of the *Salix cinerea-Betula pubescens-Phragmites australis* is rather similar in its canopy composition but seems to correspond to the 'fen carr' of Lambert (1951, 1965), which she saw as the natural development from the kinds of *Peucedano-Phragmitetum* in which either *Phragmites* or *Cladium* had been dominant.

It is difficult to assess the absolute rate of these kinds of primary succession but Lambert & Jennings (1951) noted that, along the Bure at least, the limits of the open Broads waters shown on the 1838-41 Tithe Maps coincided roughly with the original limits of lake muds in the deep medieval peat cuttings. This would suggest that, whether the Broads remained open prior to this time for natural reasons or not, most of the visible colonisation from swamp through, in places, to woodland, has occurred within the last 150 years. Stratigraphical analysis of the solid peats and clays nearer the valley sides (Jennings & Lambert 1951, Lambert & Jennings 1951) has revealed a picture of a simpler and more gradual succession following the final marine transgression. Here, vegetation dominated by *Phragmites* seems to have given way directly to 'fen carr', some of which was thought to have a possibly direct continuity with the original woodland cover (Lambert & Jennings 1951).

Over much of the extent of these firmer deposits, however, the long history of interference means that much of the woodland cover is probably of secondary origin. It is very difficult to extract from the diverse relationships between fen treatment and the vegetation, any coherent schemes of secondary succession attendant upon continued treatment or, now, its almost total demise. Nevertheless, a number of tentative observations can be made.

First, on the drier peats at Wicken, Godwin (1941) demonstrated a reversible relationship between sedge-dominated vegetation very similar to the *Symphytum* sub-community and a *Molinia caerulea*-dominated grassy sward based on the frequency of summer cuttings, the latter vegetation being favoured by annual cropping for litter. Comparable vegetation was also

Figure 15. Variations among sequences of swamp, tall-herb fen and woodland vegetation in open-water transition mires around lowland standing and sluggish waters.

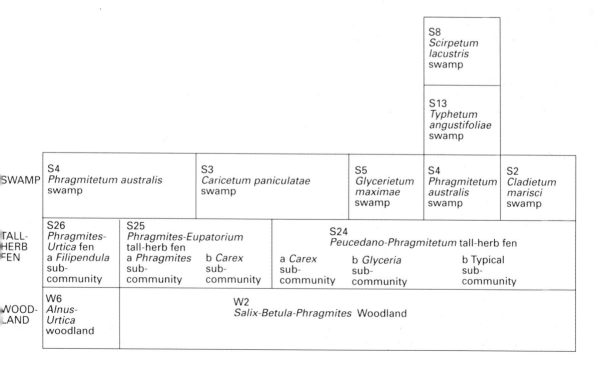

				S8 *Scirpetum lacustris* swamp		
				S13 *Typhetum angustifoliae* swamp		
SWAMP	S4 *Phragmitetum australis* swamp	S3 *Caricetum paniculatae* swamp		S5 *Glycerietum maximae* swamp	S4 *Phragmitetum australis* swamp	S2 *Cladietum marisci* swamp
TALL-HERB FEN	S26 *Phragmites-Urtica* fen a *Filipendula* sub-community	S25 *Phragmites-Eupatorium* tall-herb fen a *Phragmites* sub-community	b *Carex* sub-community	S24 *Peucedano-Phragmitetum* tall-herb fen a *Carex* sub-community	b *Glyceria* sub-community	b Typical sub-community
WOODLAND	W6 *Alnus-Urtica* woodland	W2 *Salix-Betula-Phragmites* Woodland				

shown by Lambert (1946) to be characteristic of areas along the Yare, subject to similar treatment and far from the irrigating effect of the waters. Such sustained summer mowing, especially on drier peats, seems to be one of the most effective ways of breaking the cycle of maintenance of secondary fen by converting the *Peucedano-Phragmitetum* to another vegetation type. Systematic grazing by stock, especially when combined with embankment and drainage, is probably another. On the grazed levels of the Yare, the community appears to develop into the *Holco-Juncetum* under such treatment (Lambert 1948). More drastic disturbance of drier peats may result in oxidation of the organic deposits and a release of nutrients (e.g. Haslam 1965) with an increase in nitrophilous species. Such a development is perhaps already seen within the *Glyceria* and *Symphytum* sub-communities but it may also transform the *Peucedano-Phragmitetum* into the tall-herb fens of the *Phragmites-Urtica* community or the *Epilobietum hirsutae*.

Second, how far such disruptions as these prevent the eventual re-establishment, with neglect, of the kinds of woodland that result from the primary succession is unknown. At Wicken, there seems to have been an opportunistic recolonisation of abandoned sedge-beds with the establishment of a cover largely of *Rhamnus catharticus* and *Frangula alnus* (Godwin 1936, Godwin et al. 1974), although *A. glutinosa* is regenerating in places. In the Yare fens, Lambert (1946) suggested that secondary *Alnus* woodland, though readily developing from litter-mown areas, was deficient in *Carex paniculata* as compared with primary carr. It is possible that more disturbed tall-herb fens regress to *Alnus glutinosa-Urtica dioica* woodland.

In some places, the re-development of woodland is complicated by the possibility of surface acidification of undisturbed peats which are now above the water-table. In Broadland, the *Schoenus* sub-community is associated with uncut baulks and, with a cessation of mowing, this vegetation is rapidly invaded by *Myrica gale*, first to form the *Myrica* sub-community of the *Peucedano-Phragmitetum* and then a *Myrica* scrub. This, in turn, may progress to the *Betula-Molinia* woodland (Wheeler 1980c). The recent marked expansion of *B. pubescens* on the drier peats at Wicken (Godwin et al. 1974) seems to be related to surface acidification. Indeed, it is not impossible that here and in some parts of Broadland there survive, as at Woodwalton (Poore 1956b), fragments of a pre-existing cover of acid peat that has been systematically stripped off (Wheeler 1978).

Third, the flooding of shallow peat cuttings, such as those dug quite extensively in some Broadland valleys in the mid-nineteenth century, seems to have created rather specialised conditions for the development of a somewhat different hydrosere to any of those characteristic of the broads margins themselves. Here, swamp vegetation may include stands of that distinct form of the *Galium palustre* sub-community of the *Phragmitetum* which Wheeler (1980a) termed *Cicuto-Phragmitetum* developed over a floating raft of *Typha angustifolia* remains (see also Lambert 1951). Other such flooded workings are at present occupied by rafts of the *Carex lasiocarpa* variant of the *Cicuta* sub-community of the *Peucedano-Phragmitetum* and it is possible that the two are serally related (Wheeler 1980c). Within areas containing stands of either or both of these vegetation types there are sometimes small islands of the *Sphagnum* sub-community of the *Salix cinerea-Betula pubescens-Phragmites australis* woodland often picked out by the prominence of the bright green fronds of *Dryopteris cristata* around the margins (Wheeler 1978). At least some of these seem to represent the development of ombrotrophic nuclei on floating rafts of peat which are kept out of reach of flooding with the calcareous and nutrient-rich broads waters (Wheeler 1978, 1980c, 1983; Giller 1982). Although such mires seem to have formed within the nearby Fens (e.g. Godwin & Clifford 1938, Poore 1956b; see also Godwin & Turner 1933, Walker 1970), their formation in Broadland at the present time seems to be strongly associated with artificial habitats. Even the large island at the north end of Barton Broad, which has a particularly extensive stand of this kind of woodland (Wheeler 1978), is known to have been piled around the edge and infilled with slushy dredgings (Lambert & Jennings 1965).

The vegetation of these shallow peat diggings is often further complicated by their isolation from the main paths of water movement through the broads and dykes and by the possible lateral seepage of water at the junctions of the flood-plain mire with the valley sides. It is in such positions that the *Cicuta* sub-community can occur in striking mosaics with a carpet of the *Carex rostrata-Calliergon* fen. These mosaics, which Wheeler (1978, 1980a) included within the *Peucedano-Phragmitetum* as the sub-community *caricetosum* but which are fully described in this scheme under the *Carex-Calliergon* fen, are the locus of a number of Broadland rarities, including *Liparis loeselii*, *Anagallis tenella*, *Drosera anglica*, *Parnassia palustris* and *Hypericum elodes*. The detailed environmental relationships of this kind of vegetation and its possible role in any fen succession are unknown.

Distribution

The most extensive tracts of the community occur in the middle reaches of the Yare, Bure, Ant and Thurne valleys in Broadland, east Norfolk. The *Symphytum* sub-community has been recorded from Wicken and Woodwalton Fens in Cambridgeshire and rather similar vegetation has been described from Catcott Heath in Somerset and the Ham fens in Kent.

Even in those areas which have been designated as nature reserves, the ongoing processes of terrestrialisation and recolonisation by shrubs and trees are reducing the extent of the *Peucedano-Phragmitetum* or modifying its floristics. Although a patchwork of fen and scrub may be valuable in some respects (for breeding or migrant passerines, for example: see Fuller 1982), the generally open character of former areas of mowing marsh has, in many places, been lost and some species characteristic of regularly managed fen have declined or become extinct (e.g. Wheeler 1978).

Affinities
Despite the traditional concentration on the richer fen vegetation of Broadland to the neglect of less striking communities elsewhere, the strong internal cohesion of the *Peucedano-Phragmitetum* argues powerfully for its treatment as a discrete unit with its natural centre of distribution in that region. As defined here, the community is very much as characterised by Wheeler (1975, 1978, 1980*a*) and, as such, it includes vegetation which, in some Continental schemes (e.g. Krausch 1964, Westhoff & den Held 1969), would be considered as especially rich forms of more broadly based Phragmition and Magnocaricion communities. Other European classifications (e.g. van Donselaar 1961, Korneck 1963, Segal & Westhoff 1969 in Westhoff & den Held 1969), have, however, recognised parallel vegetation types to the *Peucedano-Phragmitetum*, although they have dif-fered as to whether the affiliations of such communities are with the Magnocaricion (the view favoured by Wheeler) or the Cicution virosae.

In this scheme, the *Peucedano-Phragmitetum* lacks two of the sub-communities which Wheeler (1980*a*) distinguished. His *caricetosum* is here considered better placed within the *Carex rostrata-Calliergon* fen and his *arrhenatheretosum* is treated as part of the *Phragmites-Urtica* fen. By contrast, Wheeler's *Caricetum paniculatae peucedanetosum* (Wheeler 1978, 1980*a*) is taken into the *Peucedano-Phragmitetum* to effect a consistent separation between swamps and fens.

The community shows a complex pattern of affinities with other types of fen and with swamp, woodland and fen meadow vegetation. It is closely related in a general way to the *Phragmites-Eupatorium* fen which, though it lacks some of the rich-fen species, has a similarly varied medley of dominants and it grades through the *Glyceria* and *Symphytum* sub-communities to the *Phragmites-Urtica* fen. The range of possible helophyte dominants in the *Peucedano-Phragmitetum* provides a clear link with a variety of swamp communities and the sedge/rush layer continues to be a prominent part of some fen meadow vegetation. The community grades floristically to some woodlands, notably through the *Myrica* sub-community to the *Betula-Molinia* woodland and, in a more general way, to the woodlands with mixtures of *Alnus glutinosa*, *Salix cinerea* and *Betula pubescens*.

Floristic table S24

	a	b	c	d	e	f	g	24
Phragmites australis	V (3–7)	V (1–7)	V (1–7)	V (1–7)	V (1–7)	V (1–9)	V (1–7)	V (1–9)
Galium palustre	V (2–5)	V (1–3)	IV (1–3)	V (1–3)	V (1–3)	V (1–3)	V (1–3)	V (1–5)
Lysimachia vulgaris	IV (1–4)	V (1–3)	V (1–3)	IV (1–3)	IV (1–3)	IV (1–3)	V (1–3)	V (1–4)
Peucedanum palustre	V (1–5)	III (1–3)	III (1–5)	V (1–5)	IV (1–3)	V (1–3)	V (1–3)	V (1–5)
Eupatorium cannabinum	V (2–7)	IV (1–3)	V (1–3)	IV (1–3)	IV (1–3)	IV (1–3)	III (1–3)	IV (1–7)
Lythrum salicaria	IV (1–5)	IV (1–3)	III (1–3)	III (1–3)	IV (1–3)	IV (1–3)	IV (1–3)	IV (1–5)
Juncus subnodulosus	I (2–3)	III (1–5)	IV (1–5)	V (1–7)	V (1–5)	V (1–5)	IV (1–3)	IV (1–7)
Calliergon cuspidatum	IV (1–2)	III (1–3)	IV (1–5)	III (1–5)	V (1–3)	III (1–5)	II (1–3)	IV (1–5)
Calamagrostis canescens	II (1–3)	III (1–3)	V (1–7)	IV (1–5)	III (1–3)	II (1–3)	IV (1–7)	IV (1–7)
Filipendula ulmaria	II (1–4)	V (1–5)	V (1–5)	IV (1–5)	I (1–3)	II (1–3)	V (1–3)	IV (1–5)
Mentha aquatica	II (1)	III (1–3)	IV (1–3)	III (1–3)	V (1–3)	V (1–3)	II (1–3)	IV (1–3)
Cladium mariscus	I (1)	III (1–3)	III (1–9)	II (1–7)	IV (1–7)	V (1–7)	V (1–7)	IV (1–9)
Carex paniculata	V (5–9)	V (1–5)	I (1–3)	I (1–3)	I (1–3)		I (1–3)	II (1–9)
Scutellaria galericulata	IV (1–4)	IV (1–3)		III (1–3)	II (1–3)	II (1–3)	II (1–3)	II (1–4)
Brachythecium rutabulum	III (1–2)	III (1–3)	III (1–3)	I (1–3)	I (1–3)			II (1–3)
Solanum dulcamara	III (1–3)	III (1–3)	I (1–3)	I (1–3)	II (1–3)	I (1–3)	I (1–3)	II (1–3)
Impatiens capensis	IV (2–6)	II (1–3)		I (1–3)	I (1–3)			I (1–6)
Lemna minor	III (1–3)							I (1–3)
Scrophularia aquatica	II (1–4)							I (1–4)
Sparganium erectum	I (1–2)							I (1–2)
Glyceria maxima	I (1–3)	V (1–9)						I (1–9)
Epilobium hirsutum	III (2–8)	V (1–5)		I (1–3)	I (1–3)		I (1–3)	I (1–8)
Myosotis scorpioides	I (1–2)	IV (1–3)			I (1–3)	I (7)		I (1–7)
Carex riparia	I (2–3)	IV (1–3)			I (1–3)	I (1–3)		I (1–3)
Vicia cracca		IV (1–3)	I (1–3)					I (1–3)
Caltha palustris		IV (1–3)		I (1–3)	II (1–3)		I (1–3)	I (1–3)
Equisetum palustre		III (1–3)		I (1–3)			I (1–3)	I (1–3)
Typha latifolia		III (1–3)						I (1–3)
Carex acutiformis		IV (1–5)	V (1–3)	I (1–3)		II (1–3)	I (1–3)	II (1–5)
Thalictrum flavum		IV (1–3)	V (1–5)		I (1–3)			II (1–5)

Species	1	2	3	4	5	6	7	8	9
Calystegia sepium	II (1–3)	IV (1–5)	IV (1–3)	II (1–3)	II (1–3)	II (1–3)	III (1–3)	II (1–3)	II (1–5)
Lathyrus palustris	I (2)	IV (1–5)	III (1–3)	I (1–3)		I (1–3)	I (1–3)		II (1–5)
Symphytum officinale			IV (1–3)						I (1–3)
Phalaris arundinacea			III (1–5)		I (1–3)	I (1–3)		I (1–3)	I (1–5)
Galium uliginosum		I (1–3)	III (1–3)					I (1–3)	I (1–3)
Calamagrostis epigejos			II (1–7)						I (1–7)
Cicuta virosa	II (1–3)			I (1–3)	V (1–3)	I (1–3)			I (1–3)
Ranunculus lingua	II (1–4)	I (1–3)		I (1–3)	IV (1–3)	I (1–3)	I (1–3)		II (1–3)
Carex pseudocyperus	I (2–4)			II (1–3)	IV (1–3)	II (1–3)			I (1–4)
Berula erecta				II (1–3)	IV (1–3)	II (1–3)	II (1–3)		II (1–4)
Sium latifolium				I (1–3)	IV (1–3)	I (1–3)			I (1–3)
Typha angustifolia	II (2–4)			I (1–3)	III (1–3)	I (1–3)	I (1–3)		I (1–4)
Carex lasiocarpa					III (1–7)				I (1–7)
Carex appropinquata				I (1–3)	II (1–3)				I (1–3)
Pedicularis palustris				I (1–3)	II (5)	I (1–3)			I (1–5)
Menyanthes trifoliata		I (1–3)		I (1–3)	II (1–5)				I (1–5)
Equisetum fluviatile		I (1–3)		I (1–3)	II (1–3)				I (1–3)
Calliergon giganteum				I (1–3)	II (1–3)				I (1–3)
Riccardia pinguis					I (1–3)				I (1–3)
Carex lepidocarpa					I (1–3)				I (1–3)
Schoenus nigricans					I (1–3)	IV (1–3)			I (1–3)
Oenanthe lachenalii						IV (1–3)			I (1–3)
Ranunculus flammula					I (1–3)	III (1–3)		I (1–3)	I (1–3)
Samolus valerandi					I (1–3)	II (1–3)	I (1–3)		I (1–3)
Cirsium dissectum						I (1–3)	I (1–3)		I (1–3)
Myrica gale				I (1–3)	II (1–3)		I (1–3)	V (1–5)	II (1–5)
Salix repens			II (1–3)	I (1–3)	I (1–3)	I (1–3)	II (1–3)	IV (1–3)	II (1–3)
Betula pubescens sapling					I (1–3)		III (1–3)	III (1–3)	I (1–3)
Potentilla erecta						I (1–3)		I (1–3)	I (1–3)
Valeriana officinalis	I (2–2)	V (1–3)	II (1–3)	V (1–3)	I (1–3)	IV (1–3)	III (1–3)	III (1–3)	III (1–3)
Iris pseudacorus	II (1–2)	IV (1–3)	V (1–3)	IV (1–3)	IV (1–3)	III (1–3)		II (1–3)	III (1–3)
Lycopus europaeus	III (1–3)	III (1–3)	I (1–3)	IV (1–3)	IV (1–3)	IV (1–3)	III (1–3)	I (1–3)	III (1–3)

Floristic table S24 (*cont.*)

	a	b	c	d	e	f	g	24
Salix cinerea saplings	IV (2–7)	III (1–5)	III (1–3)	II (1–5)	III (1–3)	II (1–3)	IV (1–3)	III (1–7)
Thelypteris palustris	V (3–8)	II (1–3)		IV (1–5)	II (1–3)	I (1–3)	IV (1–3)	III (1–8)
Alnus glutinosa sapling	V (1–5)	II (1–3)		II (1–3)	I (1–3)	III (1–3)	II (1–3)	II (1–5)
Angelica sylvestris		IV (1–3)	III (1–3)	III (1–3)		I (1–3)	II (1–3)	II (1–3)
Carex elata	I (2)	IV (1–3)	I (1–3)	II (1–3)	IV (1–5)	IV (1–5)		II (1–5)
Cirsium palustre		II (1–3)	IV (1–3)	III (1–3)	I (1–3)	I (1–3)		II (1–3)
Rumex hydrolapathum	II (1–4)	III (1–3)	I (1–3)	IV (1–3)	II (1–3)	I (1–3)	II (1–3)	II (1–4)
Campylium stellatum			IV (1–5)	II (1–3)	IV (1–3)	III (1–3)		II (1–5)
Epilobium palustre		III (1–3)		II (1–3)	II (1–3)	I (1–3)	II (1–3)	II (1–3)
Potentilla palustris	I (2)			IV (1–5)	IV (1–3)	II (1–3)	III (1–3)	II (1–5)
Hydrocotyle vulgaris			I (1–3)	III (1–5)	IV (1–3)	V (1–3)	II (1–3)	II (1–5)
Oenanthe fistulosa				II (1–3)	IV (1–3)	I (1–3)	I (1–3)	II (1–3)
Cardamine pratensis	I (1)	II (1–3)		II (1–3)	II (1–3)	II (1–3)	I (1–3)	I (1–3)
Lotus uliginosus			I (1–3)	III (1–3)	I (1–3)	I (1–3)	I (1–3)	I (1–3)
Agrostis stolonifera			I (1–3)	II (1–3)	I (1–3)	III (1–3)		I (1–3)
Lychnis flos-cuculi			I (1–3)	II (1–3)	II (1–3)	II (1–3)		I (1–3)
Stellaria palustris	I (1–2)			II (1–3)	III (1–3)	I (1–3)		I (1–3)
Plagiomnium rostratum			II (1–3)	I (1–3)	I (1–3)	I (1–3)		I (1–3)
Carex panicea			III (1–3)		I (1–3)	II (1–3)	I (1–3)	I (1–3)
Molinia caerulea			II (1–3)		I (1–3)	I (1–3)	II (1–3)	I (1–3)
Myosotis laxa caespitosa				II (1–3)	I (1–3)	I (1–3)	I (1–3)	I (1–3)
Urtica dioica	I (1–2)		I (1–3)	I (1–3)				I (1–3)
Rubus fruticosus agg.			I (1–3)	I (1–3)			I (1–3)	I (1–3)
Epilobium parviflorum			I (1–3)	I (1–3)	I (1–3)		I (1–3)	I (1–3)
Rhinanthus minor			I (1–5)	I (1–3)	I (1–3)			I (1–5)
Poa trivialis		I (1–3)	I (1–3)	I (1–3)				I (1–3)
Holcus lanatus			I (1–3)	I (1–3)		I (1–3)		I (1–3)
Bryum pseudotriquetrum			I (1–3)		I (1–3)	I (1–3)		I (1–3)
Campylium elodes			I (1–3)		I (1–3)	I (1–3)		I (1–3)
Rhizomnium punctatum				I (1–3)	I (1–3)	I (1–3)		I (1–3)
Dactylorhiza incarnata			I (1–3)		I (1–3)			I (1–3)
Dactylorhiza majalis praetermissa				I (1–3)	I (1–3)	I (1–3)	I (1–3)	I (1–3)

	a	b	c	d	e	f	g	24
Plagiomnium elatum					I (1–3)			I (1–3)
Fissidens adianthoides		I (1–3)						I (1–3)
Frangula alnus sapling		I (1–3)				I (1–3)		I (1–3)
Cirsium arvense		I (1–3)						I (1–3)
Hypericum tetrapterum			I (1–3)					I (1–3)
Lathyrus pratensis			I (1–3)					I (1–3)
Valeriana dioica			I (1–3)	I (1–3)			I (1–3)	I (1–3)
Carex diandra				I (1–3)	I (2)			I (1–3)
Epipactis palustris				I (1–3)				I (1–3)
Succisa pratensis				I (1–3)	I (1–3)			I (1–3)
Number of samples	12	11	22	17	21	12	12	107
Number of species/sample	21 (14–32)	25 (17–35)	23 (17–34)	28 (19–39)	21 (14–36)	20 (14–29)		

a   *Carex paniculata* sub-community
b   *Glyceria maxima* sub-community
c   *Symphytum officinale* sub-community
d   Typical sub-community
e   *Cicuta virosa* sub-community
f   *Schoenus nigricans* sub-community
g   *Myrica gale* sub-community
24 *Peucedano-Phragmitetum* (total)

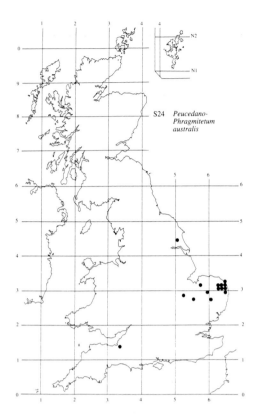

S24   *Peucedano-*
      Phragmitetum
      australis

S25
Phragmites australis-Eupatorium cannabinum tall-herb fen

Synonymy

Fen Formation Tansley 1911 *p.p.*; *Scirpo-Phragmitetum* Koch 1926 *p.p.*; Valley fen communities Bellamy & Rose 1961 *p.p.*; *Phragmites* valley fen Haslam 1965; *Angelico-Phragmitetum australis* Wheeler 1980a, not *sensu* Ratcliffe & Hattey 1982.

Constant species

Eupatorium cannabinum, *Galium palustre*, *Phragmites australis*.

Rare species

Thelypteris palustris.

Physiognomy

The most consistent feature of the rather variable vegetation included in this community is the prominence of tall herbaceous dicotyledons among which *Eupatorium cannabinum*, *Angelica sylvestris*, *Lythrum salicaria*, *Cirsium palustre*, *Valeriana officinalis*, *Iris pseudacorus*, *Filipendula ulmaria* and *Epilobium hirsutum* are the most frequent throughout. Those other tall herbs so characteristic of the *Peucedano-Phragmitetum*, *Peucedanum palustre* and *Lysimachia vulgaris*, are rare and the patchy dominance of more nutrient-demanding species typical of the *Phragmites-Urtica* fen is uncommon. Here, a variety of monocotyledons may be dominant. *Phragmites australis* is constant throughout and is often the most abundant species, but, in some stands, *Carex paniculata* or *Cladium mariscus* dominate and, more rarely, *Carex elata*, *C. riparia* or *C. acuta* may attain prominence.

Beneath, there is often some *Juncus subnodulosus*, although the amount of this is very variable, and a variety of small herbs, among which *Mentha aquatica* and *Caltha palustris* are the most frequent. *Galium palustre* is a constant sprawler with, less commonly, *Vicia cracca*, *Solanum dulcamara* and *Calystegia sepium*. Bryophytes are rarely abundant but there is occasionally a little *Calliergon cuspidatum* and *Brachythecium rutabu-*

lum. In some sub-communities, saplings of *Salix cinerea* or *Myrica gale* bushes are conspicuous.

Sub-communities

***Phragmites australis* sub-community:** *Angelico-Phragmitetum typicum* and *juncetosum subnodulosi* Wheeler 1980a. Here, *Phragmites* is usually the dominant, forming a canopy 1–2 m tall, although *E. cannabinum* and *E. hirsutum* can both be locally abundant. Denser stands of reed are rather species-poor but where the cover is more open there are small amounts of the tall and short herbaceous dicotyledons of the community and frequently a little *J. subnodulosus*. Here, too, *Lychnis floscuculi*, *Equisetum fluviatile* and *E. palustre* occurs occasionally and sprawlers are sometimes conspicuous with *Vicia cracca*, *Calystegia sepium*, *Galium uliginosum* and *Lotus uliginosus*. *Calliergon cuspidatum* and *Brachythecium rutabulum* may be prominent over litter and damp patches of bare substrate.

***Carex paniculata* sub-community:** *Caricetum paniculatae* Wangerin 1916 *p.p.*; *Angelico-Phragmitetum caricetosum paniculatae* Wheeler 1980a. The general floristics of this sub-community are very similar to the above but, although *Phragmites* remains constant, it is usually present in smaller amounts and the vegetation is dominated by the bulky tussocks of *C. paniculata*, sometimes over 1.5 m tall. *Carex acuta*, *Scrophularia aquatica* and *Phalaris arundinacea* are slightly preferential here and there are very frequently some saplings of *Salix cinerea* on the tussock tops.

***Cladium mariscus* sub-community:** *Cladietum marisci* (Allorge 1922) Zöbrist 1935 *p.p.*; *Cladium-Phragmites* consociation Wheeler 1975 *p.p.* Some of the taller herbaceous dicotyledons, notably *Cirsium palustre*, *Epilobium hirsutum*, *Angelica sylvestris* and *Valeriana officinalis*, are rare or absent here and the vegetation is usually strongly dominated by mixtures of *Cladium* and *Phrag-*

mites, sometimes with *Carex elata*. *J. subnodulosus* is constant and sometimes abundant and there is occasionally a little *Scutellaria galericulata* and *Myrica gale*. A variety of species typical of the *Peucedano-Phragmitetum* occur at low frequency, e.g. *Thelypteris palustris*, *Berula erecta*, *Oenanthe lachenalii* and *Pedicularis palustris*. Again, *Salix cinerea* saplings are frequent.

Habitat

The *Phragmites-Eupatorium* fen is most characteristic of moderately eutrophic situations where mineral or organic soils are irrigated and frequently waterlogged by usually calcareous and base-rich waters. It is typically a community of valley mires developed alongside small lowland rivers with catchments of calcareous bedrocks or superficial deposits but it also occurs in some extensive flood-plain mires and open-water transitions. It can be found, too, in spring and basin mires where more oligotrophic fen peats have been subject to moderate disturbance. Although some stands are found in sites with a history of fen treatment, the community is not maintained by mowing and is generally ungrazed.

The particular habitat feature associated with this kind of fen seems to be the moderate level of nutrient enrichment. In the most usual situation in valley mires, nutrient levels are renewed by the constant throughput of water and the occasional flooding of the alluvial terraces with the deposition of allochthonous sediments. The *Phragmites* and *Carex paniculata* sub-communities are especially characteristic of the periodically waterlogged silty fen peats or humose alluvial soils which occur on such terraces. In Breckland, where both sub-communities, and especially the former, have been described from the valley mires along the middle reaches of the Wissey, Little Ouse, Thet and Lark, Haslam (1965) noted that it was the larger amounts of available phosphate in the more substantial inorganic fraction that most clearly distinguished the soils under these vegetation types from those carrying the *Schoeno-Juncetum subnodulosi* in the headwater spring mires. Where the *Phragmites-Eupatorium* fen occurred in the latter type of mire, as the *Phragmites* or *Cladium* sub-communities, it was confined to areas where there had been some amelioration of the naturally more oligotrophic conditions: where the peat had become drier on the surface and oxidised, alongside the silt-carrying main streams, around more nutrient-rich springs, along ditches or over areas which had been dug for peat long before (see also Bellamy & Rose 1961). Similar features may mark the community's occasional occurrences in basin mires.

Zonation and succession

The *Phragmites-Eupatorium* fen occurs typically in small, often linear, stands in complexes of herbaceous and woody vegetation on mires which are now often sharply marked off from surrounding agricultural land. In more intact valley mires, the *Phragmites* and *Carex paniculata* sub-communities may form a zone along the river terraces, sometimes grading in open water to *Phragmitetum australis* swamp. On drier ground there may be a transition to the *Salix cinerea-Betula pubescens-Phragmites australis* woodland or *Alnus glutinosa-Carex paniculata* woodland. This probably represents a fairly natural succession on alluvial deposits in these calcareous and fairly eutrophic valley mires and Haslam (1965) adduced some documentary evidence that *S. cinerea* carr had succeeded *Phragmites*-dominated vegetation in 50 years in a Breckland site. Frequently, however, such successions have been disturbed and zonations are more complex (Figure 16). Peat digging and mowing for litter are known to have occurred in some valley mires and the abandonment of these practices, together with subsequent interference such as periodic burning, channel dredging, the construction of embankments and the drainage of surrounding land, is often marked now by a patchwork of the *Phragmites-Eupatorium* fen with more eutrophic communities such as the tall-herb vegetation of the *Phragmites-Urtica* fen and the *Epilobietum hirsutae* (e.g. Haslam 1965, Ratcliffe & Hattey 1982) and the *Alnus glutinosa-Urtica dioica* woodland over the dry and disturbed alluvium.

In some valley mires, there is a natural marginal transition from the community to the *Schoeno-Juncetum* along a spring-line (Haslam 1965, Wheeler 1980c). Often, though, the accessible edges of valley mires have been open to stock and grazing has obscured any zonations. In such situations, well described from the East Kent fens (Rose 1950) and the Gordano Valley in Somerset (Willis & Jefferies 1959), the *Phragmites-Eupatorium* fen may survive only around the wetter streams and dykes in a landscape of fen-meadow. Similar patterns can be seen on some disturbed spring mires, such as those described from the Little Ouse–Waveney watershed on the Norfolk–Suffolk border (Bellamy & Rose 1961, Haslam 1965).

In more eutrophic calcareous topogenous mires, the community can occur as a zone between swamps and woodland, although here, too, patterns may be confused by disturbance and truncated above by sharp transitions to agricultural land. Some open-water transitions and flood-plain mires have a belt of the *Phragmites* or *Carex paniculata* sub-communities between the *Phragmitetum* or *Caricetum paniculatae* and *Alnus* woodland, as in certain of the Shropshire Meres (Sinker 1962, Meres Report 1980). In smaller basin mires, such as some of those in Anglesey, the *Cladium* sub-community can occur as a zone grading in open water to the *Cladietum marisci* or the *Caricetum elatae* or in a mosaic with the *Schoeno-Juncetum* and fen-meadow vegetation.

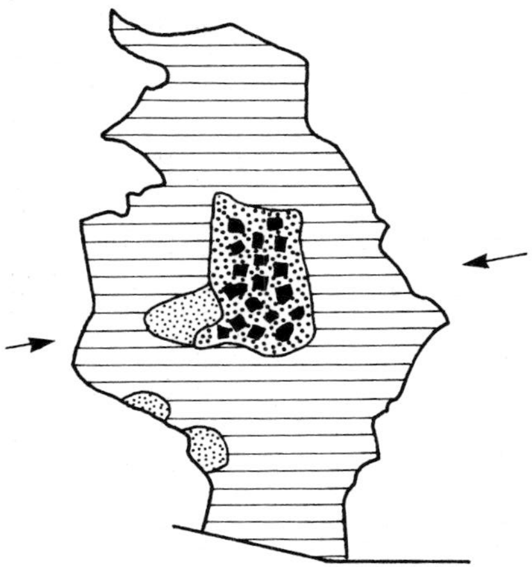

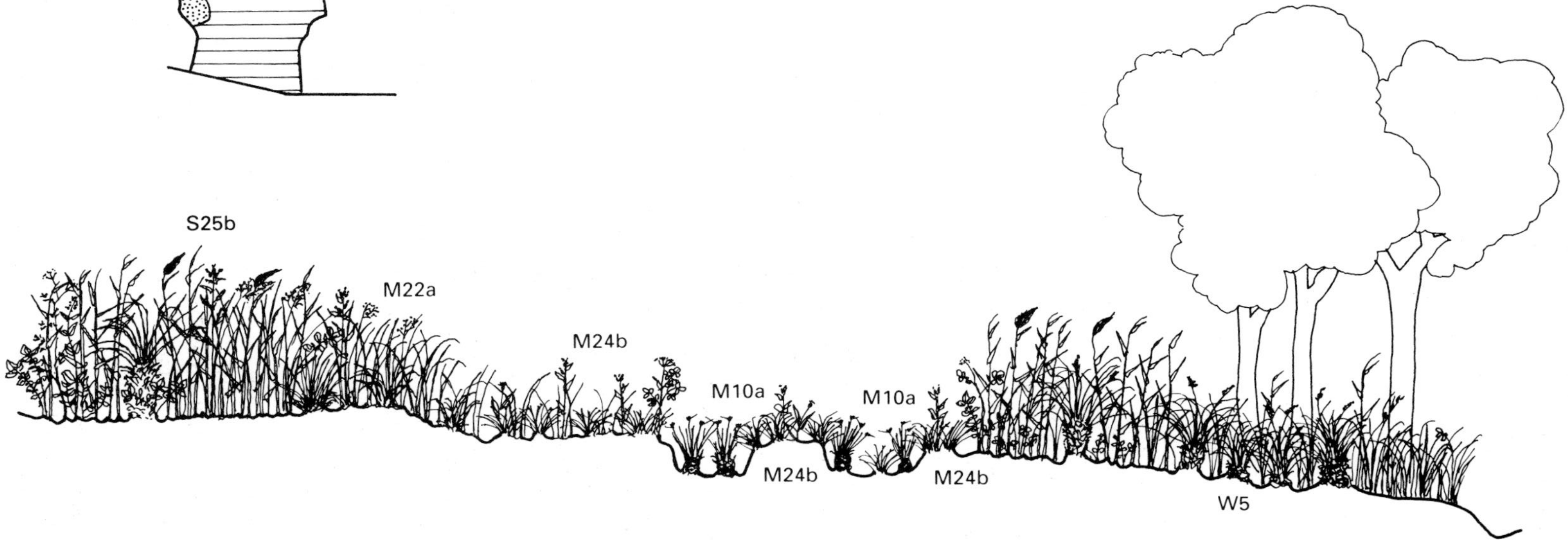

Figure 16. Map and cross-section showing pattern of fen, fen-meadow and woodland vegetation at Market Weston Fen, Suffolk (by kind permission of Suffolk Wildlife Trust).

S25b *Phragmites-Eupatorium* fen, *Carex paniculata* sub-community on intact fen peat
M22a *Juncus subnodulosus-Cirsium palustre* fen-meadow, Typical sub-community on peat with some ground-water seepage, mown annually
M24b *Cirsio-Molinietum* fen-meadow, *Eupatorium* sub-community on raised areas among old peat-cuttings
M10a *Schoeno-Juncetum* fen, *Caltha-Galium* sub-community in old peat-cuttings with ground-water seepage
W5 *Alnus-Carex paniculata* woodland on peaty alluvium with flooding from river

Distribution

The community has a widespread but scattered distribution throughout the English and Welsh lowlands and, outside Broadland, it represents the richest kind of fen vegetation. The community has not been recorded from Scotland, although some of the samples in Spence's (1964) *Carex paniculata-Angelica* sociation could perhaps be accommodated within the *Carex paniculata* sub-community.

The *Phragmites-Eupatorium* fen is especially characteristic of mires which have escaped agricultural improvement in undulating landscapes of calcareous bedrocks and drift. Particularly good stands still remain in the Breck river valleys, although some of the sites included in Haslam's (1965) survey have since been afforested. The *Carex paniculata* sub-community is well represented on the narrow alluvial flood-plains of some rivers on the Hampshire Chalk. The *Cladium* sub-community includes some of the more species-poor vegetation in the Broadland flood-plains and around the Anglesey basin mires.

Affinities

In Wheeler's (1980a) scheme, the *Angelico-Phragmitetum* was a somewhat ill-defined group which accommodated fen vegetation lacking the full range of *Peucedano-Phragmitetum* species. However, as part of a wider spectrum of both species-rich and species-poor fens, the *Phragmites-Eupatorium* community emerges here as a more integrated equivalent to this vegetation type and one which seems to occupy a fairly natural position between the *Peucedano-Phragmitetum* and the *Galium* sub-community of the *Phragmitetum*. As such it includes communities of English lowland fen vegetation described from outside Broadland (e.g. Tansley 1911, Sinker 1962, Haslam 1965, Meres Report 1980) and is equivalent to some moderately rich communities of Continental Europe.

Floristic table S25

	a	b	c	25
Phragmites australis	V (1–10)	IV (3–7)	IV (1–9)	IV (1–10)
Eupatorium cannabinum	IV (1–5)	IV (1–3)	IV (1–4)	IV (1–5)
Galium palustre	III (1–5)	IV (1–3)	III (1–3)	IV (1–5)
Cirsium palustre	III (1–2)	IV (1–3)	I (1–3)	II (1–3)
Epilobium hirsutum	III (1–5)	III (1–3)		II (1–5)
Angelica sylvestris	III (1–3)	IV (1–3)	I (1–3)	III (1–3)
Valeriana officinalis	II (1–3)	II (1–3)		II (1–3)
Vicia cracca	II (1–3)	II (1–3)		II (1–3)
Caltha palustris	II (1–3)	II (1–3)	I (1–3)	II (1–3)
Myosotis scorpioides	I (1–3)	I (1–3)		I (1–3)
Urtica dioica	I (1–3)	I (1–3)		I (1–3)
Galium uliginosum	II (1–3)	I (1–3)		I (1–3)
Calystegia sepium	II (1–3)	I (1–3)		I (1–3)
Brachythecium rutabulum	II (1–3)		I (1–5)	I (1–5)
Lychnis flos-cuculi	II (1–3)		I (1–3)	I (1–3)
Lotus uliginosus	II (1–3)			I (1–3)
Equisetum fluviatile	II (1–3)			I (1–3)
Carex riparia	I (1–3)			I (1–3)
Typha angustifolia	I (5)			I (5)
Potentilla palustris	I (2)			I (2)
Lophocolea bidentata	I (1)			I (1)
Arrhenatherum elatius	I (1)			I (1)
Athyrium filix-femina	I (2)			I (2)
Rumex crispus	I (1)			I (1)
Lathyrus pratensis	I (1)			I (1)
Galium aparine	I (1)			I (1)

Carex paniculata	I (3)	V (1–5)	I (1–3)	II (1–5)
Salix cinerea sapling	I (1–5)	IV (1–4)	III (1–4)	II (1–5)
Carex acuta	I (1–3)	II (1–5)		I (1–5)
Scrophularia aquatica	I (1–3)	II (1–3)		I (1–3)
Phalaris arundinacea		II (1–3)		I (1–3)
Juncus subnodulosus	III (1–7)	I (1–3)	V (1–4)	III (1–7)
Cladium mariscus	I (2)		V (3–8)	II (2–8)
Carex elata			II (1–9)	I (1–9)
Myrica gale			II (1–4)	I (1–4)
Scutellaria galericulata	I (2)		II (1–3)	I (1–3)
Berula erecta			I (1–4)	I (1–4)
Thelypteris palustris			I (1–3)	I (1–3)
Oenanthe lachenalii			I (1–3)	I (1–3)
Campylium stellatum			I (1–4)	I (1–4)
Valeriana dioica			I (1–3)	I (1–3)
Pedicularis palustris			I (4)	I (4)
Rubus fruticosus agg.			I (3–4)	I (3–4)
Mentha aquatica	III (1–4)	II (1–3)	III (1–3)	III (1–4)
Lythrum salicaria	II (1–3)	III (1–3)	III (1–3)	III (1–3)
Iris pseudacorus	II (1–3)	II (1–3)	II (1–3)	II (1–3)
Filipendula ulmaria	III (1–5)	II (1–3)	II (1–3)	II (1–5)
Calliergon cuspidatum	III (1–6)	I (1–3)	II (1–5)	II (1–6)
Solanum dulcamara	I (1–3)	I (1–3)	I (1–3)	I (1–3)
Equisetum palustre	II (1–3)	I (1–3)	I (1–3)	I (1–3)
Symphytum officinale	I (1–3)	I (1–3)	I (1–3)	I (1–3)
Alnus glutinosa	I (1–5)	I (3)	I (1–3)	I (1–5)
Lycopus europaeus	I (3)		I (1–3)	I (1–3)
Hydrocotyle vulgaris	I (6)		I (1–3)	I (1–6)
Menyanthes trifoliata	I (4)		I (1–5)	I (1–5)
Number of samples	16	7	26	49
Number of species/sample	10 (6–17)	12 (5–16)	11 (6–28)	11 (6–28)

a   *Phragmites australis* sub-community

b   *Carex paniculata* sub-community

c   *Cladium mariscus* sub-community

25   *Phragmites australis-Eupatorium cannabinum* tall-herb fen (total)

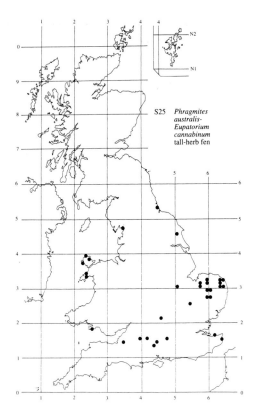

S25 *Phragmites
australis-
Eupatorium
cannabinum*
tall-herb fen

S26
Phragmites australis-Urtica dioica tall-herb fen

Synonymy

Angelico-Phragmitetum typicum sensu Ratcliffe & Hattey 1982 *p.p.*

Constant species

Phragmites australis, Urtica dioica.

Physiognomy

The tall-herb fen vegetation included in this community is very variable in its floristics and physiognomy. Apart from the two constants, *Galium aparine* is the only species that is at all frequent throughout. Moreover, although both *Phragmites* and *U. dioica* are generally abundant and often dominant, the various sub-communities are marked by the characteristically patchy local prominence of a variety of other tall dicotyledons or monocotyledons, most notably *Epilobium hirsutum, Filipendula ulmaria, Oenanthe crocata, Calystegia sepium, Solanum dulcamara, Glyceria maxima, Arrhenatherum elatius* and, less frequently, *Carex riparia* or *Phalaris arundinacea*. These form a typically chequered canopy usually 1–2 m in height which is often dense and so tangled by climbers and sprawlers as to be virtually impenetrable.

Stands are frequently species-poor and, even in more open vegetation, associates are rather varied. There are sometimes scattered plants of species characteristic of richer fens, e.g. *Lythrum salicaria, Lysimachia vulgaris, Angelica sylvestris, Cirsium palustre, Iris pseudacorus* and *Rumex hydrolapathum* and, beneath these, occasional *Pulicaria dysenterica, Equisetum fluviatile, Scutellaria galericulata* and *Silene dioica*. In other cases, there may be a grassy understorey with *Poa trivialis, Holcus lanatus* and *Dactylis glomerata*. Sprawling *Galium palustre, Lotus uliginosus* and *Rubus fruticosus* agg. sometimes add to the tangle. Bryophytes are very sparse and there are very rarely any invading shrubs or trees.

Sub-communities

Filipendula ulmaria sub-community: *Epilobium-Urtica-Galium-Filipendula* fen Haslam 1965; *Phragmites-Urtica* community Wheeler 1980*c*; *Angelico-Phragmitetum typicum sensu* Ratcliffe & Hattey 1982 *p.p.* In the often very tall, dense and species-poor vegetation of this sub-community either *Phragmites* or *U. dioica* or *F. ulmaria* can be dominant with, less frequently, a local abundance of *Eupatorium cannabinum*. Apart from *G. aparine*, no other species occurs with even occasional frequency.

Arrhenatherum elatius sub-community: *Peucedano-Phragmitetum arrhenatheretosum* Wheeler 1980*a p.p.* *Phragmites* is usually dominant here with varying amounts of *U. dioica* and *G. aparine* and, beneath, tussocks of *Arrhenatherum elatius*. Scattered plants of *Cirsium arvense* and *Heracleum sphondylium* are very characteristic of this vegetation and there is occasionally a little *Calystegia sepium* and *Angelica sylvestris*.

Oenanthe crocata sub-community. Again, *Phragmites* is normally dominant in this sub-community but there is generally only a little *U. dioica* and the most prominent associate is *Oenanthe crocata*, whose tall, robust shoots present a striking appearance among the reed. *Calystegia sepium, Vicia cracca* and, less frequently, *Lathyrus pratensis* and *Solanum dulcamara* climb among the vegetation and there are occasional scattered plants of both rich-fen tall herbs and species characteristic of disturbed and enriched habitats such as *Rumex crispus, Sonchus arvensis* and *Dipsacus fullonum*. Clumps of *Phalaris arundinacea* and tussocks of *Festuca arundinacea* can be locally abundant and there may be a patchy understorey with some *Elymus repens, Agrostis stolonifera, Carex otrubae* and *Potentilla anserina*.

Epilobium hirsutum **sub-community:** Primary *Glycerietum* and Primary *Phragmitetum* Lambert 1946 *p.p.*; *Epilobium-Urtica-Galium-Phragmites* fen Haslam 1965; *Peucedano-Phragmitetum typicum*, *Phalaris* variant Wheeler 1980a *p.p.*; *Epilobium-Filipendula* community Wheeler 1980c *p.p.*; *Angelico-Phragmitetum typicum* sensu Ratcliffe & Hattey 1982 *p.p. Phragmites* or, less frequently, *Glyceria maxima* or, rarely, *Carex riparia* is dominant here but *U. dioica* and, particularly distinctive, *Epilobium hirsutum* frequently occur in prominent patches. Where stands occur alongside woodland, *Solanum dulcamara* is sometimes abundant. There is occasionally a little *Galium aparine* and, by water margins, *Typha latifolia* may be found. Here too, scattered plants of smaller herbs can occur, e.g. *Mentha aquatica*, *Ranunculus repens*, *Myosotis scorpioides*, *Apium nodiflorum*.

Habitat

The *Phragmites-Urtica* fen is characteristic of eutrophic, circumneutral to basic water margins and mires where organic or mineral substrates are kept fairly moist throughout the year with ground-water gleying and, in some cases, winter flooding. It occurs as primary fen in some naturally more nutrient-rich open-water transitions and flood-plain mires, as along the Yare valley in Norfolk (Lambert 1946) and, particularly towards the south-west, it may form part of the richer reed-dominated vegetation that occurs above salt-marsh strandlines (Proctor 1980). Often, however, its distribution can be related to the eutrophication that may follow the drying and disturbance of fen surfaces or the contamination of ground waters by agricultural run-off, sewage or some industrial effluents. It is a common community in the moister parts of some drained and disturbed flood-plain and valley mires (e.g. Haslam 1965, Ratcliffe & Hattey 1982) and may represent the only fen vegetation that survives in much-improved lowland agricultural landscapes. It also occurs occasionally on grossly disturbed spring mires (Haslam 1965, Wheeler 1980c). It is widely distributed, though often as fragmentary strips, along ditches and canals and around ponds. The community is generally unmown but it may be accessible to grazing stock, which provide further enrichment in their dung, and it is sometimes disturbed by ditch clearing and dredging.

The distinctive feature of the habitat seems to be a certain natural or artificial balance between soil moisture and trophic state. Conditions are generally sufficiently moist for *Phragmites* to retain an overall prominence but the substrate is sufficiently dry at certain times of the year to allow nutrient-demanding perennial dicotyledons such as *U. dioica* and *E. hirsutum* to gain a hold and spread vegetatively into dense patches. There is some evidence to suggest (Haslam 1965) that, provided

moisture levels are not too high, the ability of such species to compete with *Phragmites* is increased with the greater availability of nutrients, especially perhaps of phosphate, and that, where conditions remain fairly eutrophic, they may also have the edge in drier situations. Where a suitability is roughly maintained between extremes of these variables, the characteristic tall-herb mixtures of the community seem to represent a fairly balanced state with little or no mutual suppression by the various co-dominants (Haslam 1971b).

On the flood-plains of the Yare as described by Lambert (1946), a natural stability of this kind was typical of some of the primary fen subject to irrigation with nutrient-rich waters circulating freely in response to tidal influence, although here the situation was complicated by the ability of *G. maxima* to compete with *Phragmites* in vegetation classified in this scheme in the *Epilobium* sub-community. Much of this kind of fen has now been overgrown by shrubs and trees. Elsewhere, the *Filipendula* sub-community probably represents the most natural kind of *Phragmites-Urtica* fen occurring in some less disturbed valley mires, like certain Breck fens (Haslam 1965), or in mire remnants which remain moist but where there has not been excessive eutrophication. The *Oenanthe* sub-community too may represent a natural floristic response to the enrichment by strandline detritus that occurs on upper salt-marshes flushed with fresh water, as along the Exe estuary (Proctor 1980).

In many cases, though, the balance of wetness and enrichment characteristic of the habitat has had an artificial origin. The eutrophication of lowland ground waters is now very widespread and the virtually ubiquitous distribution of the *Epilobium* sub-community throughout the agricultural lowlands and in some wet sites in derelict urban and industrial areas is one reflection of this. In other cases, it is the drainage and disturbance of mires with oxidation of organic matter and release of nutrients that marks the occurrence of the *Phragmites-Urtica* fen, especially again the *Epilobium* sub-community. In her study of Breck mires, Haslam (1965) showed how these processes could increase the trophic state of valley mire substrates and, even on the deep and oligotrophic peats of spring mires, bring about a similar encouragement of *U. dioica* and *E. hirsutum* to that obtained by the addition of NPK and PK fertilisers. In the valley mires, *Phragmites*, the natural fen dominant, was able to survive some lowering of the water-table and persist alongside these species in mixed tall-herb fens which sometimes marked old mowing marsh.

Zonation and succession

The community frequently occurs in open-water transitions and flood-plain and valley mires in association with other types of fen and passing, on wetter ground, to

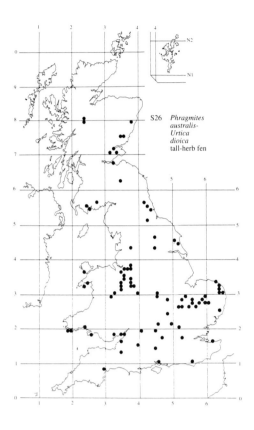

S26   *Phragmites*
       australis-
       Urtica
       dioica
       tall-herb fen

Carex rostrata-Potentilla palustris tall-herb fen
Potentillo-Caricetum rostratae   Wheeler 1980a

Synonymy

Caricetum rostratae Rübel 1912 *p.p.*; *Caricetum acuto-vesicariae comaretosum* Passarge 1955; *Carex rostrata* nodum Daniels 1978 *p.p.*

Constant species

Carex rostrata, Galium palustre, Menyanthes trifoliata, Potentilla palustris, Calliergon cordifolium/cuspidatum/giganteum.

Rare species

Carex appropinquata, C. aquatilis, C. diandra, Lysimachia thyrsiflora, Peucedanum palustre, Sium latifolium.

Physiognomy

The *Potentillo-Caricetum rostratae* comprises generally rather species-poor but somewhat varied mixtures of monocotyledons and herbaceous dicotyledons, sometimes developed as a floating mat. Among the monocotyledons, *Carex rostrata* is the most frequent species throughout but other sedges, such as *C. vesicaria, C. nigra, C. elata* and, more rarely, *C. aquatilis* and *C. appropinquata,* or *Eriophorum angustifolium* may each on occasion replace it as the dominant. In other cases, *Phragmites australis* or, less frequently, *Juncus effusus* or *J. acutiflorus* may dominate (see Sub-communities below).

A consistent feature of the community, however, is the presence of *Potentilla palustris* and *Menyanthes trifoliata* as an open or closed carpet beneath these usually taller species. They can occur in intimate mixtures or as patchy mosaics in which clumps of either attain local dominance and, scattered amongst them, is a variety of herbs, most commonly *Galium palustre* (which can sprawl extensively), *Cardamine pratensis, Epilobium palustre, Mentha aquatica, Myosotis laxa* ssp. *caespitosa, Caltha palustris, Hydrocotyle vulgaris* and *Veronica scutellata.* Taller species, such as *Angelica sylvestris, Cirsium palustre* and *Valeriana officinalis,* are sometimes conspicuous and there is a much greater

variety of these in one of the sub-communities. In wetter places, *Equisetum fluviatile* may be abundant and, here too, there may be scattered clumps of *Sparganium erectum* or *Typha latifolia. Agrostis stolonifera* with, less frequently, some *A. canina* ssp. *canina* and *Poa trivialis,* can form extensive and thick carpets, sometimes semi-submerged.

Almost always, one or more of the larger *Calliergon* spp., *C. cuspidatum, C. cordifolium* and *C. giganteum,* occur and these may form conspicuous patches over the mass of interweaving rhizomes and litter. No other bryophytes are frequent throughout but individual stands may have some *Sphagnum squarrosum* or *S. fimbriatum* or other locally abundant species.

Sub-communities

Carex rostrata-Equisetum fluviatile **sub-community:** Herbaceous Marsh Matthews 1914 *p.p.*; Mixed Fen Holdgate 1955b *p.p.*; Rich fen Sinker 1960 *p.p.*; *Carex rostrata-Acrocladium cordifolium/cuspidatum, Potentilla-Acrocladium cordifolium* and *Equisetum fluviatile-Acrocladium cordifolium* sociations Spence 1964; *Potentilla-Acrocladium* nodum Proctor 1974 *p.p.*; 'General Fen' Adam *et al.* 1975 *p.p.*; *Potentillo-Caricetum rostratae typicum* Wheeler 1980a; *Potentillo-Caricetum rostratae, Lemna minor* variant and Typical variant *p.p.* Ratcliffe & Hattey 1982; includes *Caricetum vesicariae* Br.-Bl. & Denis 1926 and *Lysimachio-Caricetum aquatilis* Neumann 1957 *sensu* Birse 1980. Here are included more species-poor stands in which *C. rostrata* is generally the most abundant monocotyledon forming a thin to dense cover of shoots 30–70 cm tall. In some stands, however, *C. vesicaria* or *C. aquatilis* can dominate and, on occasion, *P. palustris, M. trifoliata* or *E. fluviatile* (which is preferential for this sub-community) can be locally abundant. Such vegetation can have a semi-swampy appearance or develop as a floating mat and in areas of open water there may be some *Lemna minor* and various small herbs of water margins, such as

Veronica beccabunga or *Apium nodiflorum*. In other cases, *Juncus acutiflorus* or *J. effusus* may dominate. As well as the large *Calliergon* spp., there is sometimes an abundance of *Brachythecium rutabulum*, *B. rivulare*, *Rhizomnium punctatum* or *R. pseudopunctatum*. Among rare species recorded here are *Carex diandra*, *C. appropinquata* and *Lysimachia thyrsiflora*.

Lysimachia vulgaris sub-community: *Carex elata* consocies and Mixed fen associes Pearsall 1918 *p.p.*; *Potentillo-Caricetum rostratae lysimachietosum* Wheeler 1980a. *C. rostrata* remains constant here and is sometimes abundant but, more usually, the dominant is *Phragmites*, forming a tall open canopy, *Carex elata* or *C. nigra*, the latter usually in a robust tussocky form (Wheeler 1980a, 1982; Jermy *et al.* 1982), *Juncus effusus* or *Eriophorum angustifolium*. Beneath and around these, *P. palustris* and *M. trifoliata* remain prominent with their scatter of small herbs but this sub-community is marked by the high frequency and sometimes the abundance of tall dicotyledons more usually associated with rich-fen vegetation: *Lysimachia vulgaris*, *Lythrum salicaria*, *Peucedanum palustre*, *Lycopus europaeus*, *Lychnis flos-cuculi*, *Ranunculus lingua*, *Rumex hydrolapathum*, *Sium latifolium* and *Iris pseudacorus*. There is also often some *Juncus subnodulosus* and *Campylium stellatum* is a frequent component of the bryophyte layer.

Habitat

The *Potentillo-Caricetum* is most characteristic of peaty soils kept moist by mesotrophic to oligotrophic, neutral to moderately base-rich and calcareous waters. Conditions seem rarely to be stagnant. Although the community is most frequent in topogenous mires, occurring in open-water transitions and basin mires, stands commonly experience some unseasonal flooding and the community seems best developed around areas of diffuse lateral water flow near gentle inflow and outflow streams. It is also found in more obviously soligenous situations: where, for example, throughput ameliorates ombrotrophic conditions around and within some raised mires. It is a primary fen community, although it is sometimes maintained by grazing.

The soil is usually a raw peat or humic gley with a surface or water pH between 5 and 7 and dissolved calcium levels of 10–90 mg l^{-1} (Holdgate 1955b, Sinker 1960, Spence 1964, Proctor 1974, Wheeler 1983). The water-table, even in dry weather, is probably at or just below the surface for much of the time (Matthews 1914, Pearsall 1918, Holdgate 1955b, Spence 1964). In the growing season, the vegetation is thus kept sufficiently moist but free of lengthy inundation for the characteristic carpet of larger *Calliergon* spp. to develop (Clapham 1940, Segal 1966). The growth of the robust, creeping rhizomes of the understorey of *Potentilla palustris* and *Menyanthes trifoliata* and the accumulation of litter probably play an important role in raising the substrate surface to this level (Matthews 1914, Hewett 1964) and the moss carpet itself may contribute to peat formation and provide a place where smaller herbaceous species can gain a hold (Matthews 1914, Clapham 1940).

Conditions remain, however, sufficiently base-rich and calcareous to inhibit the development of ombrotrophic nuclei and the formation of a *Sphagnum* carpet such as is typical of the rather similar vegetation of the *Carex rostrata-Sphagnum squarrosum* mire. The brief periods of unseasonal flooding characteristic of many basin mires and even moderately large lakes around which the community occurs may be important here. To the north and west, where the *Potentillo-Caricetum* is most widespread, heavy summer rains can produce sudden, sometimes marked, but usually short-lived changes in water-level in such sites (e.g. Holdgate 1955b, Proctor 1974, Wheeler 1980d, Lock & Rodwell 1981). Even in summer, stands can be encountered with water-tables at any level from almost 40 cm below the surface to more than 10 cm above (Matthews 1914, Pearsall 1918, Holdgate 1955b, Spence 1964). In some cases, the vegetation can develop as a floating raft of interweaving rhizomes up to 50 cm thick over peaty or silty ooze and this may rise and fall somewhat with any changes in water-level (Lock & Rodwell 1981) but, even where the vegetation is rooted in a solid substrate, it seems tolerant of this kind of inundation. The atypical tussock form of *Carex nigra*, characteristic of some stands of the community, may indeed be a response to strong fluctuations in water-level (Wheeler 1980a, d).

Differences in the balance between nutrient levels, base-richness and calcium content may play some part in influencing the floristic variation between the sub-communities. The *Carex-Equisetum* sub-community is found over the full extent of the pH and calcium content ranges, sometimes in quite mesotrophic mires. It is especially characteristic of open-water transitions around more nutrient-rich lakes to the north and west where base-status and calcium content can be low (Spence 1964) but it also occurs in more base-rich and calcareous basin mires which can be quite oligotrophic (Proctor 1974). The *Lysimachia* sub-community is more strictly confined to this latter site type, occurring, for example, in mires developed in hollows in glacial drift, but it is sometimes found in open-water transitions (e.g. Pearsall 1918) and in artificial hollows (such as peat cuttings or duck decoy pools) in flood-plain mires (Wheeler 1980a, 1983). It is also more eastern in its distribution and climate may play some part in its floristic composition.

More accessible stands of the *Potentillo-Caricetum* are often grazed by stock which can eat shrub and tree

seedlings and which perhaps favour the dominance of
Juncus spp. that is a marked feature in some places.

Zonation and succession
In more extensive and undisturbed open-water transitions, the *Carex-Equisetum* sub-community (and, rarely
the *Lysimachia* sub-community) occurs as part of complete zonations between submerged aquatic vegetation
and woodland. Around lakes and some larger basin
mires, it is generally fronted by a zone of the *Caricetum
rostratae*, the *Equisetetum fluviatile* or, at scattered
localities, especially in Scotland, by the *Caricetum vesicariae*. Boundaries between these swamps and the
Potentillo-Caricetum are often very hazy with a gradual
increase in shallower water in *Potentilla palustris* and
Menyanthes trifoliata and a continuing prominence of
the swamp dominants within the fen. In other cases,
virtually pure stands of *P. palustris* or *M. trifoliata*
themselves may extend out into open water in front of
the community (Matthews 1914, Spence 1964, Birse
1980, Lock & Rodwell 1981). At some sites, a belt of the
Phragmitetum forms a more abrupt outer edge to the
Potentillo-Caricetum (e.g. Pearsall 1918, Lock & Rodwell 1981).

On drier ground, the community may grade, through
an open scatter of *Salix cinerea* bushes, to the *Salix-Carex rostrata* woodland in which some important
Potentillo-Caricetum species remain as part of the
understorey. There is little doubt that such sequences
represent a natural succession around more mesotrophic, sometimes base-rich and calcareous lakes and
basins to the north and west. At Esthwaite Fen in
Cumbria, it has been shown how the *Potentillo-Caricetum* has extended out behind a front of the *Phragmitetum* and been colonised from behind by *S. cinerea* over
the past seventy years (Pearsall 1918, Tansley 1939,
Pigott & Wilson 1978) and Lock & Rodwell (1981)
adduced photographic evidence to support a similar
succession at Crag Lough in Northumberland. Spence
(1964) described a replacement of *Caricetum rostratae*
by the *Potentillo-Caricetum* over fifty years on the shores
of Loch Ness in Inverness, although at other sites there
has been no perceptible change in the extent of these two
communities. The levels of nutrients may play some part
in influencing the rate of forward advance of the fen mat
in such situations.

Two things confuse this basic pattern of zonation and
succession. The first is grazing, which can set back shrub
and tree invasion where stock have access to the
community. The effect of this is starkly visible at Crag
Lough where a fence marks a sharp boundary between
open *Potentillo-Caricetum* and closed *Salix-Carex*
woodland. Grazing may also blur the boundary between
the fen vegetation on the mire surface itself and that on
the mineral or peaty soils surrounding it. Even where

there is no woodland fringe to the community, this
junction may be marked by a zone of Filipendulion
vegetation, the *Holco-Juncetum* or the *Phalaridetum
arundinaceae* at the limit of inundation or ground-water
gleying. Where such margins are grazed, there tends to
be a much more gradual transition from rush-pasture or
fen-meadow communities to the *Potentillo-Caricetum*
(e.g. Ratcliffe & Hattey 1982).

The second feature, which has more complex results,
is the variation in base-status, calcium content and
nutrient levels that is very characteristic of soligenous
areas in basins and around some raised mires. Such
differences, which can be very marked over even short
distances, are partly a function of distance from springs,
seepage lines or streams but they can also be affected by
ground-water fluctuations. The *Potentillo-Caricetum* is
often found in places where it is difficult to separate the
influence of soligenous, topogenous and ombrogenous
effects and it frequently forms part of intricate mosaics
with other mire communities. Particularly striking
examples have been described from Sunbiggin Tarn in
Cumbria (Holdgate 1955*b*) and Malham Tarn, North
Yorkshire (Sinker 1960, Proctor 1974, Adam *et al.*
1975). Here, around the weaving inflow streams, the
community occurs intimately mixed with the *Carex
rostrata-Calliergon* fen in a baffling jumble of local
differences in bryophyte distribution and dominance by
a variety of Carices. The vagaries of dispersal and
establishment may play some part in determining such
patterns but they are probably also influenced by differences in water chemistry, although these have not been
isolated (Proctor 1974, Wheeler 1980*a*, *b*, 1983). In such
sites as these, there may also be more obvious zonations
between either or both these fens and the *Pinguiculo-Caricetum dioicae* around highly calcareous springs and
to Filipendulion vegetation along junctions with
unflushed mineral soils or where banks of alluvium have
been deposited by moving waters. A patchy development of the *Salix-Carex* woodland may add to the
complexity.

In less base-rich and calcareous mires, the *Potentillo-Caricetum* may occur in mosaics with poor-fen vegetation. Small patches of *Sphagnum squarrosum* and *S.
fimbriatum* are sometimes found within stands of the
community, where, for example, the surfaces of floating
rafts are maintained at a high enough level to be free of
frequent inundation (Wheeler 1980*a*, *d*; Lock & Rodwell
1981). Where such patches coalesce, other Sphagna,
such as *S. palustre*, *S. recurvum* and *S. teres*, and
Aulacomnium palustre may appear and *Calliergon stramineum* replace *C. giganteum* and *C. cordifolium*. Such
changes mark a transition to the *Carex rostrata-Sphagnum squarrosum* community. In other places, the fading
of any nutrient enrichment in the water flowing into
basin or around raised mires is accompanied by a

gradation from the *Potentillo-Caricetum* to the much less species-rich *Carex rostrata-Sphagnum recurvum* mire, which has but scattered plants of *M. trifoliata* and *P. palustris* and a very poor representation of other herbs (e.g. Wheeler 1980*d*). Where fragments of ombrotrophic peat remain in such soligenous mires (as in the Malham fens) or where water tracks run in a well-defined lagg around raised mires, the community may pass very sharply to some form of ombrogenous bog or its derivative.

Distribution

The *Carex-Equisetum* sub-community is very much a vegetation type of the north and west with a wide distribution in open-water transitions and basin mires in Wales, northern England and Scotland. The *Lysimachia* sub-community extends the range of the community into eastern England where it occurs uncommonly in small basin mires and rarely in the flood-plain mires of Broadland (Wheeler 1980*a, d*, 1983).

Affinities

The *Potentillo-Caricetum* has affinities with a wide variety of other mire types and it is difficult to define its floristic limits exactly. First, it grades through its bryophyte layer to a variety of poor fens of the Caricion canescentis-fuscae and Rhynchosporion alliances in which *C. rostrata* and Junci remain prominent. In some

of these, there is also an understorey of *P. palustris* and *M. trifoliata* but they can generally be distinguished by the prominence of a range of Sphagna. Second, it shows clear affinities with some richer Caricion davallianae communities. Here again, *C. rostrata, P. palustris* and *M. trifoliata* may remain prominent but other sedges, notably *C. diandra, C. lasiocarpa, C. nigra* and *C. panicea*, become more important and there is a shift in the bryophyte layer to more calcicolous species. Each of the various communities involved has a fairly well defined core of distinguishing species but there is a virtually continuous spectrum of variation between them all.

If a *Potentillo-Caricetum* is defined from this range of vegetation types, there is an obvious case for retaining it as a rich Caricion canescentis-fuscae community or a poor Rhynchosporion community but here we follow Wheeler (1980*a*) in placing it within the Magnocaricion. Its links with the diverse tall dicotyledon element of rich fens are seen only in the *Lysimachia* sub-community where it approaches the vegetation of some stands of the *Peucedano-Phragmitetum* but many of its dominants occur as important swamp species in the Phragmitetea and it has strong developmental relationships with some of these. Indeed, some authorities (e.g. Birse 1980) have seen vegetation of this kind as part of association(s) which also include species-poor swamps.

Floristic table S27

	a	b	27
Carex rostrata	V (1–9)	V (1–9)	V (1–9)
Galium palustre	V (1–5)	V (1–4)	V (1–5)
Potentilla palustris	IV (1–8)	V (1–9)	V (1–9)
Menyanthes trifoliata	IV (1–9)	IV (1–9)	IV (1–9)
Equisetum fluviatile	IV (1–8)	II (1–5)	III (1–8)
Juncus acutiflorus	II (1–6)		I (1–6)
Ranunculus flammula	II (1–3)		I (1–3)
Carex vesicaria	I (8–9)		I (8–9)
Carex aquatilis	I (8–9)		I (8–9)
Veronica beccabunga	I (1–5)		I (1–5)
Apium nodiflorum	I (4–5)		I (4–5)
Brachythecium rutabulum	I (1–6)		I (1–6)
Brachythecium rivulare	I (1–4)		I (1–4)
Stellaria alsine	I (1–3)		I (1–3)
Rhizomnium punctatum	I (1–5)		I (1–5)
Rhizomnium pseudopunctatum	I (1–4)		I (1–4)
Potamogeton polygonifolius	I (2)		I (2)
Callitriche stagnalis	I (3)		I (3)
Hypericum elodes	I (7)		I (7)

Floristic table S27 (*cont.*)

	a	b	27
Ranunculus repens	I (3)		I (3)
Senecio aquaticus	I (1)		I (1)
Carex diandra	I (3)		I (3)
Lysimachia thyrsiflora	I (4)		I (4)
Cardamine amara	I (3)		I (3)
Epipactis palustris	I (1)		I (1)
Angelica sylvestris	II (1–3)	IV (1–4)	III (1–4)
Eriophorum angustifolium	I (1–3)	IV (3–9)	II (1–9)
Phragmites australis	I (5)	IV (4–9)	II (4–9)
Lythrum salicaria	I (1)	IV (1–9)	II (1–9)
Lysimachia vulgaris		IV (1–5)	II (1–5)
Carex nigra	I (3–5)	III (1–9)	II (1–9)
Ranunculus lingua	I (1–3)	III (1–6)	II (1–6)
Lycopus europaeus	I (1–2)	III (1–4)	II (1–4)
Iris pseudacorus	I (1–4)	III (1–3)	II (1–4)
Carex elata	I (5)	III (3–8)	II (3–8)
Lychnis flos-cuculi	I (1)	III (1–3)	II (1–3)
Campylium stellatum	I (1)	III (1–4)	II (1–4)
Juncus subnodulosus		III (1–7)	II (1–7)
Peucedanum palustre		II (1–6)	I (1–6)
Rumex hydrolapathum		II (1–3)	I (1–3)
Sium latifolium		II (1–3)	I (1–3)
Sphagnum fimbriatum		I (1–5)	I (1–5)
Dryopteris cristata		I (1)	I (1)
Cardamine pratensis	III (1–4)	IV (1–3)	III (1–4)
Agrostis stolonifera	III (1–5)	III (1–9)	III (1–9)
Juncus effusus	II (2–8)	III (2–9)	III (2–9)
Epilobium palustre	II (1–6)	III (1–3)	II (1–6)
Mentha aquatica	II (3–8)	III (1–5)	II (1–8)
Calliergon cuspidatum	II (1–5)	III (1–3)	II (1–5)
Calliergon cordifolium	II (2–9)	II (1–2)	II (1–9)
Myosotis laxa caespitosa	II (1–4)	II (1)	II (1–4)
Caltha palustris	II (1–4)	II (1)	II (1–4)
Hydrocotyle vulgaris	II (1–4)	II (2–5)	II (1–5)
Calliergon giganteum	II (1–7)	I (1)	I (1–7)
Agrostis canina canina	II (2–5)	I (1–3)	I (1–5)
Veronica scutellata	II (1–4)	I (2)	I (1–4)
Lemna minor	II (2–3)	I (1)	I (1–3)
Stellaria palustris	I (1–3)	II (1–5)	I (1–5)
Juncus articulatus	I (1–6)	II (2–4)	I (1–6)
Scutellaria galericulata	I (1)	II (1)	I (1)
Cirsium palustre	I (1)	II (1–3)	I (1–3)
Valeriana officinalis	I (1)	II (1–3)	I (1–3)
Myosotis scorpioides	I (2–6)	I (1)	I (1–6)
Solanum dulcamara	I (3–4)	I (1–2)	I (1–4)

Polygonum amphibium	I (3–6)	I (2–4)	I (2–6)
Sparganium erectum	I (2–5)	I (1–3)	I (1–5)
Filipendula ulmaria	I (1–2)	I (1–2)	I (1–2)
Carex paniculata	I (1–3)	I (1–4)	I (1–4)
Poa trivialis	I (1–2)	I (1)	I (1–2)
Lotus uliginosus	I (2–3)	I (1)	I (1–3)
Holcus lanatus	I (1–6)	I (1–4)	I (1-4)
Eleocharis palustris	I (1–3)	I (1)	I (1–3)
Typha latifolia	I (2–4)	I (1–2)	I (1–4)
Sphagnum squarrosum	I (2)	I (1–5)	I (1–5)
Salix cinerea sapling	I (3)	I (1–4)	I (1–4)
Carex appropinquata	I (8)	I (1)	I (1–8)
Drepanocladus fluitans	I (4)	I (1)	I (1–4)
Equisetum palustre	I (2)	I (1)	I (1–2)
Chiloscyphus polyanthos	I (1)	I (2)	I (1–2)
Bryum pseudotriquetrum	I (1)	I (1)	I (1)
Cratoneuron filicinum	I (1)	I (1)	I (1)
Lophocolea bidentata s.l.	I (1)	I (1)	I (1)
Number of samples	197	23	220
Number of species/sample	10 (6–16)		

a   *Carex rostrata-Equisetum fluviatile* sub-community
b   *Lysimachia vulgaris* sub-community
27   *Potentillo-Caricetum rostratae* (total)

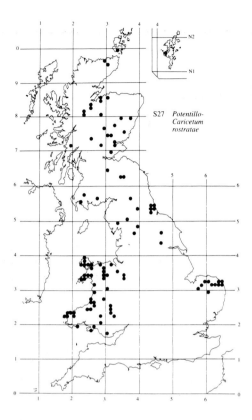

S27 *Potentillo-Caricetum rostratae*

S28
Phalaris arundinacea tall-herb fen
Phalaridetum arundinaceae   Libbert 1931

Synonymy
Reed-swamp Pallis 1911 *p.p.*; *Phalaris arundinacea consocies* Pearsall 1918; *Phalaris arundinacea* fen Tansley 1939; *Phalaris arundinacea* society and *Phalaris-Filipendula* sociation Spence 1964; Sociatie van *Phalaris arundinacea* Westhoff & den Held 1969; Tall grass washlands Ratcliffe 1977 *p.p.*; *Phalaris* nodum Daniels 1978; *Angelico-Phragmitetum typicum sensu* Ratcliffe & Hattey 1982 *p.p.*

Constant species
Phalaris arundinacea.

Physiognomy
The *Phalaridetum arundinaceae* comprises vegetation in which *Phalaris arundinacea* is dominant, forming an often dense canopy, usually 1–1.5 m tall. The vegetation is almost always species-poor and, although certain species attain prominence in some sub-communities, no associate is frequent throughout.

Sub-communities

***Phalaris arundinacea* sub-community.** Here are included very species-poor stands overwhelmingly dominated by *P. arundinacea*. Some are pure; others have scattered associates as a very sparse understorey, sometimes small herbs or helophytes of water margins, in other cases, tall herbs and sprawlers of fens or salt-marsh plants.

***Epilobium hirsutum-Urtica dioica* sub-community.** *P. arundinacea* generally remains dominant here but the canopy is more varied with mixtures of *Epilobium hirsutum* and *Urtica dioica* and, more rarely, *Glyceria maxima*. There is sometimes a little *Galium aparine* and very occasionally some *Eurhynchium praelongum*.

***Elymus repens-Holcus lanatus* sub-community.** An often shorter but usually still closed canopy of *P. arundinacea* here has beneath it a grassy understorey with one or more of *Elymus repens*, *Holcus lanatus*, *Poa trivialis* and *Deschampsia cespitosa*. There are sometimes scattered plants of *Ranunculus repens* and *Cirsium arvense* but other species are rare.

Habitat
The *Phalaridetum* is typical of the margins of fluctuating, circumneutral and mesotrophic to eutrophic waters, both standing and running. Although it can be found on organic soils, it is more characteristic of mineral substrates, from fine clays to coarse gravels. It is common in open-water transitions around ponds and lakes of all sizes and also occurs around reservoirs, flooded clay and gravel pits, in some flood-plain and basin mires and, rarely, on salt-marshes. It is widespread, too, along periodically flooded dykes and by rivers, even swift and spatey hill streams, and may occur patchily on river shoals. The vegetation may be grazed by stock or wildfowl.

Although *P. arundinacea* can be found growing in 40 cm or more of water, it will not tolerate permanent flooding and stands of the community have a summer water-table that is below the surface for most of the season (Spence 1964, Haslam 1978). Often, however, there is some unseasonal fluctuation in water-level and the community seems to thrive on strongly gleyed soils. *P. arundinacea* has a firmly anchored, creeping rhizome (Hubbard 1984) and is resistant to erosion and to the turbulence of flood-waters (Haslam 1978). The community can thus maintain itself even where inundation is sudden and substantial. In general, the grassier vegetation of the *Elymus-Holcus* sub-community is more characteristic of drier situations which may have little flooding, even in winter.

The community is often found in sites of alluvial deposition (e.g. Pearsall 1918) but *P. arundinacea* does not seem to need especially high nutrient levels. Indeed, in more eutrophic situations, its dominance may be challenged by more nutrient-demanding tall herbs, as in the *Epilobium-Urtica* sub-community, which is the most frequent form of the vegetation around pools and along streams enriched by agricultural, industrial or domestic effluents.

P. arundinacea is a palatable grass which can yield a large and continuing amount of succulent herbage throughout the growing season (Hubbard 1984). Stands alongside lowland streams may be accessible to stock and, in washlands, the community may form part of the patchwork of vegetation that is summer-grazed by cattle, sheep and horses. Here, too, it may provide some grazing for the large populations of overwintering herbivorous wildfowl (Ratcliffe 1977, Ogilvie 1978, Fuller 1982) and a valuable nesting habitat for birds such as snipe and redshank.

Zonation and succession

The community is often found as the terminating vegetation type around open-water transitions, marking the normal upper limit of water-level fluctuation. On the margins of lakes and pools it may pass, towards open water, directly to swamps such as the *Phragmitetum australis*, the *Glycerietum maximae*, the *Typhetum latifoliae* or the *Sparganietum erecti*, or there may be an intervening zone of the *Phragmites-Eupatorium* fen. Around enriched waters or in disturbed mires, the *Epilobium-Urtica* sub-community frequently forms an upper fringe to a patchwork of the *Phragmites-Urtica* fen. On the draw-down zones around reservoirs with silty shores, the community often passes sharply to a sequence of inundation communities. In its rare occurrences on salt-marches, the *Phalaridetum* normally forms a discontinuous and narrow fringe at the extreme upper limit, although in some estuaries it may occur further down-marsh above pioneer *Scirpetum maritimi* (e.g. Gilham 1957a).

Similar zonations to those found around open standing waters, though often much condensed, are also characteristic of the margins of lowland water-courses with silty or clayey banks. On the more stony banks

typical of faster-flowing streams over resistant rocks, the *Phalaridetum* is sometimes the only tall herbaceous fen vegetation, passing more or less directly to the moving open waters. On periodically inundated river shoals, clumps of the community may occur in mosaics with the *Festuca arundinacea-Agrostis stolonifera* community.

Stands of the community may grade, at their upper edge, to *Alnus glutinosa-Urtica dioica* woodland or, by streams in upland margins, to *Alnus glutinosa-Fraxinus excelsior-Lysimachia nemorum* woodland, and occasionally clumps of *P. arundinacea* can occur in the field layers of these woodlands. Often, however, the upper margins of the *Phalaridetum* are sharply marked off from neighbouring pasture land. On drier soils where there is no grazing, there may be a more gradual transition through the *Elymus-Holcus* sub-community to some kind of *Arrhenatheretum* or to the *Deschampsia cespitosa-Holcus lanatus* grassland. On winter-flooded washlands, the *Phalaridetum* can pass, away from the less heavily grazed margins of dykes, to a mosaic of the *Glycerietum fluitantis* and the *Agrostis stolonifera-Alopecurus geniculatus* inundation grassland.

Shrub and tree seedlings and saplings are rare in the *Phalaridetum* and no observational data is available as to succession but it seems most likely that the community progresses normally to the *Alnus-Urtica* woodland, or on the upland margins of the north and west, to the *Alnus-Fraxinus-Lysimachia* woodland.

Distribution

The *Phalaridetum* is a widespread and common community throughout the British lowlands and on the upland margins.

Affinities

The community is most closely related to the tall-herb fen vegetation of more eutrophic open-water transitions and mires and it has traditionally been placed in the Magnocaricion.

Floristic table S28

	a	b	c	28
Phalaris arundinacea	V (5–10)	V (5–10)	V (9–10)	V (5–10)
Juncus effusus	I (1–5)	I (2)		I (1–5)
Galium palustre	I (2–4)			I (2–4)
Callitriche stagnalis	I (2–3)			I (2–3)
Myosotis scorpioides	I (3–4)			I (3–4)
Eleocharis palustris	I (3–4)			I (3–4)
Typha angustifolia	I (3)			I (3)
Puccinellia maritima	I (4)			I (4)
Epilobium hirsutum	I (2)	III (1–8)	I (3)	II (1–8)
Urtica dioica		III (1–5)		II (1–5)
Galium aparine	I (1)	II (1–4)		I (1–4)
Glyceria maxima		I (4–6)		I (4–6)
Eurhynchium praelongum		I (3–4)		I (3–4)
Elymus repens		I (1–3)	III (1–3)	I (1–3)
Holcus lanatus			III (2–4)	I (2–4)
Cirsium arvense	I (2)	I (2–3)	II (1–3)	I (1–3)
Ranunculus repens	I (1)	I (4)	II (1–4)	I (1–4)
Poa trivialis	I (2–3)	I (1)	II (1–4)	I (1–4)
Deschampsia cespitosa			II (3–4)	I (3–4)
Cirsium palustre			I (2–3)	I (2–3)
Solanum dulcamara	I (1–3)	I (1–4)	I (3)	I (1–4)
Agrostis stolonifera	I (1–2)	I (3–4)	I (4)	I (1–4)
Filipendula ulmaria	I (1–4)	I (1–3)		I (1–4)
Angelica sylvestris	I (1–2)	I (1)		I (1–2)
Mentha aquatica	I (1–2)	I (3)		I (1–3)
Calystegia sepium	I (2)	I (1–4)		I (1–4)
Oenanthe crocata	I (1–4)	I (1)		I (1–4)
Nasturtium officinale	I (4–5)	I (3–4)		I (3–5)
Equisetum fluviatile	I (4–5)	I (1–3)		I (1–5)
Atriplex prostrata	I (3)	I (3)		I (3)
Rubus fruticosus agg.	I (2–3)		I (4)	I (2–4)
Number of samples	31	20	10	61
Number of species/sample	4 (1–10)	6 (3–16)	7 (3–11)	5 (1–16)
Vegetation height (cm)	124 (25–200)	114 (40–200)	100 (30–170)	106 (25–200)
Vegetation cover (%)	94 (50–100)	98 (90–100)	100	95 (50–100)

a *Phalaris arundinacea* sub-community
b *Epilobium hirsutum-Urtica dioica* sub-community
c *Elymus repens-Holcus lanatus* sub-community
28 *Phalaridetum arundinaceae* (total)

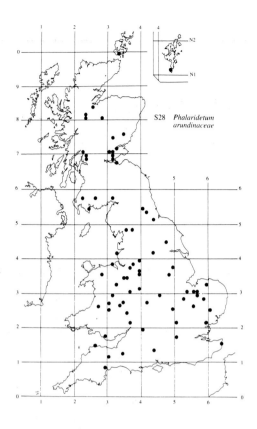

S28 *Phalaridetum
arundinaceae*

INDEX OF SYNONYMS TO AQUATIC COMMUNITIES, SWAMPS AND TALL-HERB FENS

The vegetation types are listed alphabetically, then by date of ascription of the name, with the code number of the equivalent NVC community thereafter. The NVC communities themselves are included in the list with a bold code.

Acoretum calami Schulz 1941 S15

Acorus calamus swamp **S15**

Angelico-Phragmitetum australis Wheeler 1980*a* S25

Angelico-Phragmitetum caricetosum paniculatae Wheeler 1980*a* S25

Angelico-Phragmitetum caricetosum ripariae Ratcliffe & Hattey 1982 S6

Angelico-Phragmitetum juncetosum subnodulosi Wheeler 1980*a* S25

Angelico-Phragmitetum typicum Wheeler 1980*a* S25

Angelico-Phragmitetum typicum *sensu* Ratcliffe & Hattey 1982 S26, S28

Anthropogenic *Phragmitetum, Calamagrostis* and *Juncus* fens Lambert 1951 S24

Association à *Scirpus lacustris* et *Glyceria aquatica* Allorge 1922 S5

Association of *Carex elata* Pigott & Wilson 1978 S1

Association of *Phragmites australis* and *Schoenoplectus lacustris* Pigott & Wilson 1978 S4, S8

Brackish water communities Birks 1973 S21

Bure Valley fen Pallis 1911 S24

Callitriche stagnalis community **A16**

Carex acutiformis fen Lambert 1951 S7

Carex acutiformis stands Meres Survey 1980 S7

Carex acutiformis swamp **S7**

Carex ampullacea consocies Matthews 1914 S9

Carex elata associations Holdgate 1955*b* S1

Carex elata consocies Pearsall 1918 S1, S27

Carex elata headwater fen community Haslam 1965 S1

Carex elata swamp **S1**

Carex otrubae swamp **S18**

Carex paniculata swamp **S3**

Carex paniculata swamp Poore & Walker 1959 S3

Carex paniculata swamp Sinker 1962 S3

Carex paniculata-Angelica sylvestris sociation Spence 1964 S3

Carex paniculata-Rubus fruticosus community Ratcliffe & Hattey 1982 S3

Carex pseudocyperus swamp **S17**

Carex riparia swamp **S6**

Carex rostrata 'reedswamps' Holdgate 1955*b* S9

Carex rostrata nodum Daniels 1978 S9, S27

Carex rostrata reedswamps Proctor 1974 S9

Carex rostrata sociations Spence 1964 S9

Carex rostrata swamp **S9**

Carex rostrata-Acrocladium cordifolium/cuspidatum sociation Spence 1964 S27

Carex rostrata-Menyanthes sociation Spence 1964 S9

Carex rostrata-Menyanthes trifoliata Association Birks 1973 S9

Carex rostrata-Potentilla palustris tall-herb fen **S27**

Carex vesicaria swamp **S11**

Carex vesicaria-Veronica scutellata sociation Spence 1964 S11

Caricetum Pearsall 1920 S11

Caricetum actutiformo-ripariae Soó (1927) 1930 S6, S7

Caricetum acutiformis Sauer 1937 S7

Caricetum acutiformo-paniculatae Vl. & van Zinderen Bakker 1942 S6, S7, S24

Caricetum acuto-vesicariae comaretosum Passarge 1955 S27

Caricetum elatae Koch 1926 S1

Caricetum elatae cladietosum Libbert 1932 S2

Caricetum inflatae Tansley 1939 S9

Caricetum otrubae Mirza 1978 S18

Caricetum paniculatae Tansley 1939 S3

Caricetum paniculatae   Wangerin 1916   S3, S25

Caricetum paniculatae caricetosum acutiformis   Segal & Westhoff 1969   S24

Caricetum paniculatae peucedanetosum   Wheeler 1978   S24

Caricetum paniculatae thelypteridetosum   Segal & Westhoff 1969   S24

Caricetum paniculatae typicum   Wheeler 1980*a*   S3

Caricetum ripariae   Soó 1928   S6

Caricetum rostratae   Birse 1980   S9

Caricetum rostratae   Rübel 1912   S9, S27

Caricetum vesicariae   Br.-Bl. & Denis 1926   S11, S27

Caricetum vulpinae   R.Tx. 1947   S18

Catcott Heath fen   Willis 1967   S24

Ceratophylletum demersi   Hild 1956   A5

Ceratophylletum demersi   (Pop 1962) den Hartog & Segal 1964   A5

Ceratophylletum submersi   den Hartog & Segal 1964   A6

Ceratophyllum demersum community   **A5**

Ceratophyllum demersum vegetation   Ellis 1939   A5

Ceratophyllum demersum vegetation   Lambert & Jennings 1951   A5

Ceratophyllum demersum vegetation   Lambert 1965   A5

Ceratophyllum demersum vegetation   Pallis 1911   A5

Ceratophyllum submersum community   **A6**

Ceratophyllum submersum sociatie   den Hartog 1963   A6

Ceratophyllum-Stratiotes nodum   Wheeler & Giller 1982   A4

Chara-Myriophyllum alterniflorum sociation   Spence 1964   A13, A14

Cicuto-Caricetum pseudocyperi comaretosum   Van Donselaar 1961   S24

Cicuto-Phragmitetum   Wheeler 1978   S4

Cladietum   Tansley 1939   S2

Cladietum marisci   (Allorge 1922) Zöbrist 1935   S25

Cladietum marisci   Zöbrist 1933 *emend.* Pfeiffer 1961   S2

Cladietum marisci phragmitetosum   Westhoff & Segal 1969   S2

Cladietum marisci scorpidietosum   Segal & Westhoff 1969   S2

Cladietum marisci   *sensu* Krausch 1964   S24

Cladietum marisci   *sensu* Westhoff & den Held 1969   S24

Cladietum marisci typicum   Krausch 1964   S2

Cladietum marisci typicum   Wheeler 1980*a*   S2

Cladietum marisci utricularietosum and *caricetum lasiocarpae*   Wheeler 1980*a*   S2

Cladium anthropogenic fen   Lambert 1951   S24

Cladium mariscus primary fen   Lambert 1951   S2

Cladium mariscus reed-swamp   Conway 1942   S2

Cladium mariscus society   Poore & Walker 1959   S2

Cladium mariscus stands   Holdgate 1955*b*   S2

Cladium mariscus swamp   Ratcliffe & Hattey 1982   S2

Cladium mariscus swamp and sedge-beds   **S2**

Cladium mariscus-Myrica gale sociation   Spence 1964   S2

Cladium mariscus-Rubus fruticosus-Myrica gale community   Ratcliffe & Hattey 1982   S2

Cladium mariscus-Utricularia intermedia nodum   Ivimey-Cook & Proctor 1966   S2

Cladium sociation   Wheeler 1980*a*   S2

Cladium-Carex elata community   Wheeler 1980*a*   S2

Cladium-Phragmites consociation   Wheeler 1975*a*   S25

Cladium-Thelypteris community   Wheeler 1980*a*   S2

Eleocharis palustris community   Birse 1980   S19

Eleocharis palustris consocies   Pearsall 1918   S19

Eleocharis palustris-Littorella sociation   Spence 1964   S19

Eleocharis palustris swamp   **S19**

Eleocharis palustris-Agrostis stolonifera nodum   Adam 1981   S19

Eleocharitetum acicularis   Koch 1926   A22

Eleocharitetum multicaulis   Tüxen 1937   A22

Eleocharitetum palustris   Schennikow 1919   S19

Elodea canadensis community   **A15**

Elodea society   Spence 1964   A15

Elodeetum canadensis   (Pignatti 1953) Passarge 1964   A15

Elodetum   Matthews 1914   A15

Elodetum   Tansley 1939   A15

Epilobium-Filipendula community   Wheeler 1980*c*   S26

Epilobium-Urtica-Galium-Filipendula fen   Haslam 1965   S26

Epilobium-Urtica-Galium-Phragmites fen   Haslam 1965   S26

Equisetetum fluviatile   Steffen 1931 *emend.* Wilczek 1935   S10

Equisetum fluviatile reedswamp   Tansley 1939   S10

Equisetum fluviatile swamp   **S10**

Equisetum fluviatile-Acrocladium cordifolium sociation   Spence 1964   S27

Equisetum limosum reedswamp   Rankin 1911   S10

Eriocaulo-Lobelietum   Br.-Bl. & Tx. 1952 *sensu* Birse 1984   A22

Esthwaite reedswamp   Pearsall 1918   S12

Fen Association   Pallis 1911   S24

Fen Formation   Tansley 1911   S25

Floating-leaf association   Pallis 1911   A8, A9

Floating-leaf vegetation   Butcher 1933   A9

Floating-leaf vegetation   Tansley 1939   A9, A10

Floating-leaved vegetation   Tansley 1911   A10

Floating-leaved vegetation   West 1910   A10

Gemeenschap van *Acorus calamus* en *Iris*

pseudacorus Olivier & Segal 1963 S15

'General Fen' Adam *et al.* 1975 S27

Glyceria aquatica reedswamp Pallis 1911 S5

Glyceria aquatica riparian reedswamp Tansley 1911 S5

Glyceria fluitans water-margin vegetation **S22**

Glyceria maxima communities Lambert 1947*c* S5

Glyceria maxima floating reedswamp Lambert 1946 S5

Glyceria maxima riparian zone Lambert 1947*c* S5

Glyceria maxima sociation Wheeler 1975*a* S5

Glyceria maxima stagnant reedswamp Lambert 1947*c* S5

Glyceria maxima swamp **S5**

Glyceria reedswamp Tansley 1939 S5

Glyceria society Spence 1964 S5

Glycerieto-Typhetum latifoliae Neuhäusl 1959 S5, S12

Glycerietum aquaticae Tansley 1911 S5

Glycerietum fluitantis Wilczek 1935 S22

Glycerietum maximae Tansley 1939 S5

Glycerietum maximae (Nowinski 1928) Hueck 1931 emend. Krausch 1965 S5

Glycerietum maximae Hueck 1931 S24

Glycerietum maximae riparian reedswamp Butcher 1933 S5

Glycerio-Sparganion Br.-Bl. & Sissingh *apud* Boer 1942 *emend.* Segal **S23**

Halo-Scirpetum maritimi (van Langendonck 1931) Dahl & Hadač 1941 S21

Herbaceous Marsh Matthews 1914 S27

Hydrocharis morsus-ranae-Stratiotes aloides community **A4**

Hydrocharitetum morsus-ranae van Langendonck 1935 A3, A4

Hydrocharito-Stratiotetum (van Langendonck 1935) Westhoff 1946 A4

Isoeto-Lobelietum (Koch 1926) Tx. 1937, inops and *eleocharetosum* Schoof-van Pelt 1973 A22

Isoetes lacustris/setacea community **A23**

Isoetes lacustris consocies Pearsall 1918 A23

Isoetes lacustris consocies Pearsall 1921 A23

Isoetes lacustris society Spence 1964 A23

Isoeto-Lobelietum (Koch 1926) Tx. 1937 A22

Isoeto-Lobelietum (Koch 1926) Tx. 1937 *sensu* Schoof-van Pelt 1973 *p.p.* A22, A23

Isoeto-Lobelietum (Koch 1926) Tx. 1937 *sensu* Birse 1980 A22

Isoeto-Lobelietum (Koch 1926) Tx. 1937 *sensu* Birse 1984 A22, A23

Junco subnodulosi-Calamagrostietum canascentis Korneck 1963 S24

Juncus bulbosus community **A24**

Juncus fluitans consocies Pearsall 1921 A24

Juncus fluitans vegetation Pearsall 1921 A24

Juncus fluitans-Lobelia dortmanna association Spence 1964 A22

Juncus fluitans-Sphagnum subsecundum sociation Spence 1964 A24

Juncus fluitans-Utricularia sociation Spence 1964 A24

Lemna gibba community **A1**

Lemna minor community **A2**

Lemnetum gibbae Miyawaki & J.Tx. 1960 A1

Lemnetum minoris Soó 1947 A2

Lemnetum minoris (Oberdorfer 1957) Müll. & Gors 1970 A2

Linear-leaved associes Pearsall 1918 A13

Littorella uniflora-Lobelia dortmanna community **A22**

Littorella uniflora-Lobelia dortmanna Association Birks 1973 A22

Littorella-Lobelia associes Pearsall 1918 A22

Lobelia dortmanna association Spence 1964 A22

Lobelia-Littorella and *Littorella-Juncus* sociations Spence 1964 A22

Lysimachio-Caricetum aquatilis Neumann 1957 *sensu* Birse 1980 S27

Medium-slow current vegetation Butcher 1933 A15

Mixed Fen Holdgate 1955*b* S27

Mixed fen associes Pearsall 1918 S27

Moderate current vegetation Butcher 1933 A11

Moderate-moderately swift current vegetation Butcher 1933 A18

Moderately swift current vegetation Butcher 1933 A17

Mowing marsh vegetation Lambert 1948 S24

Myriophylletum alterniflori Lemée 1937 A14

Myriophyllo-Nupharetum Koch 1926 A8

Myriophyllum alterniflorum community **A14**

Myriophyllum alterniflorum consocies Pearsall 1918 A14

Myriophyllum alterniflorum sociation Spence 1964 A14

Nuphar lutea community **A8**

Nuphar lutea society Spence 1964 A8

Nuphar lutea-Lemna vegetation Tansley 1911 A8

Nymphaea alba community **A7**

Nymphaea alba consocies Pearsall 1918 A7

Nymphaea alba sociations Spence 1964 A7

Nymphaea alba ssp. *occidentalis* sociation Spence 1964 A7

Nymphaea occidentalis consocies Pearsall 1918 A7

Nymphaeetum albae Oberdorfer & Mitarb. 1967 A7

Nymphaeetum albae Tansley 1939 A7

Nymphaeetum albo-luteae Novinski 1927 A8

Nymphaeetum minoris Vollmar 1947 A7

Nymphaeetum occidentalis Tansley 1939 A7

Open *Carex rostrata* sociation Spence 1964 S9

Open carr   Pearsall 1918   S1

Open *Equisetum fluviatile* sociation   Spence
   1964   S10

Open *Littorella-Lobelia* sociation   Spence 1964   A22

Other water-margin vegetation   **S23**

Peaty moorland lochs   West 1910   A24

Peucedano-Phragmitetum arrhenatheretosum Wheeler
   1980*a*   S26

Peucedano-Phragmitetum australis   Wheeler 1978
   emend.   S24

Peucedano-Phragmitetum cicutetosum   Wheeler
   1978   S24

*Peucedano-Phragmitetum cicutetosum, Carex
   lasiocarpa* variant   Wheeler 1978   S24

Peucedano-Phragmitetum glycerietosum   Wheeler
   1980*a*   S24

Peucedano-Phragmitetum myricetosum   Wheeler
   1978   S24

Peucedano-Phragmitetum schoenetosum   Wheeler
   1978   S24

Peucedano-Phragmitetum symphytetosum   Wheeler
   1980*a*   S24

Peucedano-Phragmitetum typicum, Phalaris
   variant   Wheeler 1980*a*   S26

Peucedano-Phragmitetum typicum, typical
   variant   Wheeler 1978   S24

Pflanzengesellschaft mit *Elodea canadensis*   Solinska
   1963   A15

Phalaridetum arundinaceae   Libbert 1931   S28

Phalaris arundinacea consocies   Pearsall 1918   S28

Phalaris arundinacea fen   Tansley 1939   S28

Phalaris arundinacea society   Spence 1964   S28

Phalaris arundinacea tall-herb fen   **S28**

Phalaris nodum   Daniels 1978   S28

Phalaris-Filipendula sociation   Spence 1964   S28

Phragmites australis reedswamp, species-poor
   variant   Meres Report 1980   S4

Phragmites australis swamp and reed-beds   **S4**

Phragmites australis-Eupatorium cannabinum tall-herb
   fen   **S25**

Phragmites australis-Peucedanum palustre tall-herb
   fen   **S24**

Phragmites australis-Urtica dioica tall-herb fen   **S26**

Phragmites communis reedbeds   Proctor 1980   S4

Phragmites communis nodum   Adam 1981   S4

Phragmites communis reedswamp   Tansley 1911   S4

Phragmites communis society   Spence 1964   S4

Phragmites communis swamp sociation   Wheeler
   1978   S4

Phragmites communis-Galium palustre
   sociation   Spence 1964   S4

Phragmites communis-Littorella open
   sociation   Spence 1964   S4

Phragmites communis-Sparganium minimum open
   sociation   Spence 1964   S4

Phragmites marsh   Ranwell 1961   S4

Phragmites monodominant stands   Haslam
   1971*a*   S4

Phragmites nodum   Daniels 1978   S4

Phragmites valley fen   Haslam 1965   S25

Phragmites-Scirpus associes   Pearsall 1918   S4

Phragmites-Scirpus associes   Tansley 1939   S8

Phragmites-Urtica community   Wheeler 1980*c*   S26

Phragmitetum   Chapman 1964   S4

Phragmitetum   Matthews 1914   S4

Phragmitetum australis   (Gams 1927) Schmale
   1939   S4

Polygonum amphibium Gesellschaft   Oberdorfer
   1977   A10

Polygonum amphibium community   **A10**

Polygonum amphibium Community   Birse 1980   A10

Polygonum amphibium sociation   Spence 1964   A10

Potameto-Nupharetum   Müller & Gors 1960   A8

Potametum pectinato-perfoliati   Den Hartog & Segal
   1964   A11

Potamogeton filiformis-Chara sociation   Spence
   1964   A11

Potamogeton gramineus society   Spence 1964   A13

Potamogeton natans community   **A9**

Potamogeton natans consocies   Pearsall 1920   A9

Potamogeton natans Gesellschaft   Oberdorfer
   1977   A9

Potamogeton natans-Juncus fluitans sociation   Spence
   1964   A9

Potamogeton pectinatus community   **A12**

Potamogeton pectinatus-Gesellschaft   Oberdorfer
   1977   A12

Potamogeton pectinatus-Myriophyllum spicatum
   community   **A11**

Potamogeton perfoliatus Gesellschaft   Oberdorfer
   1977   A11

Potamogeton perfoliatus-Myriophyllum alterniflorum
   community   **A13**

Potentilla-Acrocladium nodum   Proctor 1974   S27

Potentilla-Acrocladium cordifolium sociation   Spence
   1964   S27

Potentillo-Caricetum rostratae   Wheeler 1980*a*   S27

*Potentillo-Caricetum rostratae
   lysimachietosum*   Wheeler 1980*a*   S27

Potentillo-Caricetum rostratae typicum   Wheeler
   1980*a*   S27

Potentillo-Caricetum rostratae, Lemna minor
   variant   Ratcliffe & Hattey 1982   S27

Potentillo-Caricetum rostratae, Typical
   variant   Ratcliffe & Hattey 1982   S27

Primary *Glycerietum*   Lambert 1946   S26

Primary *Glycerietum* hover fen   Lambert 1946   S24

Primary *Phragmitetum*   Lambert 1946   S24, S26

Primary tussock fen   Lambert 1951   S3, S24

Pure Sedge   Godwin & Tansley 1929   S2

Ranunculetum aquatilis   Géhu 1961   A19

Ranunculetum baudotii   Br.-Bl. 1952   A21

Glyceria fluitans   **A3**, A8, A9, A10, A11, A13, A15, A18, A19, A20, A21, A24, S4, S8, S11, S12, S13, S14, S16, S19, S21, **S22**, S23

Glyceria maxima   S4, **S5**, S13, S16, S17, S21, S23, *S24*, S26, S28

Glyceria × *pedicellata*   A19

Glyceria plicata   A8, A20, S13

Groenlandia densa   A12, S14

Halimione portulacoides   S4, S21

Heracleum sphondylium   S23, S26

Hesperis matronalis   S14

Hippuris vulgaris   A7, A8, *A11*, A13, A21, S2, S4

Holcus lanatus   S1, S4, S7, S14, S15, S17, S19, S23, S24, S26, S27, S28

Holcus mollis   S14, S15

Hottonia palustris   A4

Hydrocharis morsus-ranae   A1, **A3**, **A4**, A11, S8, S14, S16

Hydrocotyle vulgaris   S1, S4, S8, S9, S11, S12, S19, S21, *S24*, S25, S27

Hypericum elodes   S27

Hypericum tetrapterum   S24

Impatiens capensis   S24

Iris pseudacorus   S4, S5, S8, S13, S14, S17, S18, S21, S23, *S24*, S25, S26, S27

Isoetes lacustris   A13, A22, **A23**

Isoetes setacea   A13, A22, A23

Juncus acutiflorus   S1, S6, S8, S9, S12, S18, S19, S27

Juncus articulatus   A22, S4, S9, S12, S14, S19, S21, S22, S23, S27

Juncus bufonius   S15, S18, S20, S21, S23

Juncus bulbosus   *A7*, A8, A11, A13, A14, **A24**

Juncus bulbosus/kochii   A9, A22, A23, S4, S8, S9, S18, S19

Juncus conglomeratus   S8, S17

Juncus effusus   S4, S5, S6, S7, S8, S9, S11, S12, S13, S14, S15, **S17**, S18, S19, S23, S26, S27, S28

Juncus filiformis   A10

Juncus gerardi   S4, S19, S20, S21

Juncus inflexus   S7, S14, S17, S23

Juncus maritimus   S4, S21

Juncus subnodulosus   S1, S2, S4, **S24**, *S25*, S26, S27

Lathyrus palustris   *S24*

Lathyrus pratensis   S7, S24, S25, S26

Lemna gibba   **A1**, A2, **A3**, A5, A12, A15, A18, A19, A20, A21, S8, S14, S23

Lemna minor   A1, **A2**, **A3**, **A4**, *A5*, A7, A8, A9, A10, A11, A12, A13, A14, A15, A16, A17, A18, A20, A21, S3, S4, S5, S7, S8, S10, S12, S13, S14, *S15*, S16, S22, S23, S27

Lemna trisulca   A1, *A2*, **A3**, **A4**, A5, A8, A9, A11,

A12, A13, A15, A19, A20, S12, S13, S14, S16, S23

Leontodon autumnalis   S19

Limonium vulgare   S4, S21

Littorella uniflora   A8, A9, A10, A11, **A13**, A14, **A22**, A23, S9, S13, *S19*

Lobelia dortmanna   A9, A11, A13, A14, **A22**, A23, A24, S9, S19

Lolium perenne   S23

Lophocolea bidentata s.l.   S25, S27

Lotus uliginosus   S3, S4, S7, S14, S18, S24, S25, S26, S27

Luronium natans   A11

Lychnis flos-cuculi   S4, S24, S25, S26, S27

Lycopus europaeus   S1, S3, S4, S5, S6, S8, S12, S14, S15, S17, *S24*, S25, S26, S27

Lysimachia thyrsiflora   S27

Lysimachia vulgaris   S13, **S24**, S26, *S27*

Lythrum salicaria   S1, S2, S4, S5, S7, S8, S11, S12, S14, S15, S18, S23, **S24**, S25, S26, *S27*

Marchantia polymorpha   S15

Matricaria maritima   S4, S21

Mentha aquatica   A1, A8, A18, S1, S2, S3, S4, S5, S6, S7, S8, S9, *S11*, *S12*, S13, *S14*, S15, S16, S17, S18, S19, S21, S22, S23, *S24*, S25, S26, S27, S28

Menyanthes trifoliata   S1, *S2*, *S4*, S8, *S9*, *S10*, S11, S12, S13, S24, S25, **S27**

Mimulus guttatus   S15

Mnium hornum   S27

Molinia caerulea   S3, S4, S24

Myosotis laxa caespitosa   S4, S9, S12, S14, S15, S19, S24, S27

Myosotis scorpioides   A5, S3, S4, S5, S6, S11, S12, S14, S15, S17, S22, S23, *S24*, S25, S26, S27, S28

Myrica gale   S2, *S24*, S25

Myriophyllum alterniflorum   A7, A9, A11, **A13**, **A14**, A18, A19, *A22*, A23, A24, S8, S19

Myriophyllum spicatum   A2, A5, A6, A8, **A11**, A12, A13, A15, A16, A18, A19, A21, S13

Myriophyllum verticillatum   A4

Najas flexilis   A13

Nasturtium officinale   A1, A3, A4, A5, A17, A20, S5, S6, S12, S14, S22, *S23*, S28

Nitella spp.   A7, A8, A11, *A13*, A14, A23, A24

Nuphar lutea   A4, A5, A7, **A8**, A10, A11, A13, A15, S8, S14

Nuphar pumila   A7, A13

Nymphaea alba   A4, **A7**, *A8*, A9, A10, A11, A13, A15, S4, S5, S6, S8, S9, S13

Nymphoides peltata   A11

Oenanthe aquatica   A4, A11, S23

Oenanthe crocata   S3, S4, S6, S14, S17, S21, *S26*, S28

Oenanthe fistulosa   S4, S23, *S24*

Oenanthe lachenalii   S4, S20, S21, *S24*, S25
Oenanthe silaifolia   S23
Osmunda regalis   S24

Pedicularis palustris   S9, S24, S25
Peucedanum palustre   **S24**, S27
Phalaris arundinacea   A18, S4, S5, S6, S7, S11, S12, S13, *S14*, S19, S23, S24, S25, S26, S28
Phragmites australis   S1, *S2*, S3, **S4**, S5, S6, S8, S13, S14, S17, S18, S20, S21, S23, **S24**, **S25**, **S26**, *S27*
Plagiomnium elatum   S24
Plagiomnium rostratum   S24
Plantago major   S18, S23
Plantago maritima   S4, S19, S21
Poa annua   S18
Poa pratensis   S18, S23
Poa trivialis   S4, S5, S6, S7, S15, S17, S19, S22, S23, S24, S26, S27, S28
Pohlia proligera   S15
Polygonum amphibium   A3, **A4**, A8, **A10**, A11, A12, A13, A15, A17, A19, A20, S5, S6, S8, S9, S10, S11, S12, S13, S14, S15, S19, S23, S27
Polygonum aviculare   S7, S15, S23
Polygonum bistorta   S8
Polygonum hydropiper   A17, S6, S7, S10, S23
Polygonum persicaria   S12, S14, S15, S19, S23
Potamogeton alpinus   A11, A13
Potamogeton berchtoldii   A7, A8, A11, A13, A23, A24
Potamogeton coloratus   A5
Potamogeton compressus   A11
Potamogeton crispus   A3, A4, A8, A11, A13, A15, A17, A23
Potamogeton filiformis   *A11*, A13, A14
Potamogeton friesii   A11, A13
Potamogeton gramineus   A10, A11, **A13**, A14, A24
Potamogeton gramineus nodosus   S4
Potamogeton gramineus rutilus   A13
Potamogeton lucens   A2, A11, A13, A15, S8
Potamogeton natans   A2, *A7*, A8, **A9**, A10, A11, A13, A14, A15, A19, A20, A24, S4, S8, S9, S13, S14, S15, S19
Potamogeton × *nitens*   A11, A13
Potamogeton obtusifolius   A4, A7, A8, A10, A11, A13, A23, A24
Potamogeton pectinatus   A1, A2, A5, **A6**, A8, **A11**, **A12**, A13, A15, *A16*, A17, A19, A21, S13
Potamogeton perfoliatus   A8, A10, A11, A12, **A13**, A15, A18, A23, S8
Potamogeton polygonifolius   *A7*, A11, A13, A24, S4, S8, S9, S12, S19, S27
Potamogeton praelongus   A11, A13
Potamogeton pusillus   A8, *A11*, A13
Potamogeton trichoides   A11
Potamogeton × *zizii*   A13

Potentilla anserina   S19, S20, *S21*, S23, S26
Potentilla erecta   S24
Potentilla palustris   A4, S1, S2, S3, S4, S8, S9, *S10*, S11, S12, *S24*, S25, **S27**
Potentilla reptans   S18
Puccinellia distans   S20, S21
Puccinellia maritima   *S4*, S20, S21, S28
Pulicaria dysenterica   S26

Ranunculus acris   S5, S7, S11, S19, S26
Ranunculus baudotii   **A6**, A11, A13, **A21**
Ranunculus circinatus   A3, *A5*, A8, A11, A15, A20
Ranunculus ficaria   S5
Ranunculus flammula   A22, S4, S8, S9, S10, S11, S12, S13, S14, S19, S21, S24, S27
Ranunculus fluitans   **A18**, A19, S23
Ranunculus hederaceus   A8, A13
Ranunculus lingua   A4, S1, S4, S17, S22, *S24*, S27
Ranunculus omiophyllus   A11
Ranunculus peltatus   A2, A11, A12, A13, A14, **A20**, S23
Ranunculus penicillatus pseudofluitans   A15, **A17**
Ranunculus repens   S5, S7, S12, S15, S16, S18, S19, S23, S26, S27, S28
Ranunculus sceleratus   A1, A20, S4, S12, S20, S21, *S23*
Ranunculus trichophyllus   A1, A6, A8, A9, A11, A13, A20, S23
Rhinanthus minor   S24
Rhizomnium pseudopunctatum   S27
Rhizomnium punctatum   S24, S27
Rhytidiadelphus squarrosus   S1
Riccardia multifida   S2
Riccia fluitans   A2, A8, S5, S15
Ricciocarpus natans   A2
Rorippa islandica   S10, S23
Rorippa sylvestris   S23
Rosa dumalis   S23
Rubus fruticosus agg.   S3, S4, S6, S17, S18, S24, S25, S26, S28
Rumex acetosa   S1, S3, S11, S15
Rumex conglomeratus   S21
Rumex crispus   S4, S5, S6, S7, S12, S14, S17, S18, S21, S22, S23, S25, S26
Rumex hydrolapathum   S3, S4, S5, S6, S12, S13, S15, S17, S18, S20, *S24*, S26, S27
Rumex obtusifolius   S5, S17, S23
Ruppia maritima   A12, A21, S20, S21
Ruppia spiralis   A11

Sagittaria sagittifolia   A4, A5, A8, S8, **S16**
Salicornia dolichostachya   S4, S21
Salix cinerea   S1, S2, S4, S15, *S24*, *S25*, S27
Salix repens   *S24*
Samolus valerandi   S21, S24

Schoenus nigricans   *S24*
Scirpus fluitans   A13, A14, A22, A23, A24
Scirpus lacustris lacustris   A9, **S8**, S12, S16
Scirpus lacustris tabernaemontani   S12, S13, **S20**, S21
Scirpus maritimus   S4, S18, S20, **S21**
Scorpidium scorpioides   S2, S19
Scrophularia auriculata   S17, S24, S25
Scutellaria galericulata   S3, S6, S7, S11, S14, S17, *S24*, S25, S26, S27
Senecio aquaticus   S27
Senecio jacobaea   S23
Silene dioica   S26
Sium latifolium   A4, S4, S18, *S24*, S27
Solanum dulcamara   S1, S2, S3, S4, S5, S6, S7, S8, S10, S12, S13, S14, S15, S17, S18, S20, S22, S23, S24, S25, S26, S27, S28
Solanum nigrum   S8
Sonchus arvensis   S4, S18, S21, S26
Sonchus asper   S23
Sparganium angustifolium   A7, A9, A13, A23, A24, S8
Sparganium emersum   A4, A5, A8, A11, A13, A15, A21, S16
Sparganium erectum   A4, A9, A20, S3, S4, S5, S6, S7, *S8*, S12, **S14**, S15, S16, **S17**, S18, S22, S26, S27
Sparganium minimum   A7, A8, A9, A13, A24, S4
Spartina anglica   S20, S21
Spergularia marina   S21
Spergularia media   S21
Sphagnum fimbriatum   S27
Sphagnum recurvum   S4
Sphagnum squarrosum   S27
Sphagnum subnitens   S2
Spirodela polyrhiza   A1, **A3**, A8, A15, S8, S16
Stachys palustris   S8, S16, S26
Stellaria alsine   S7, S15, S17, S23, S27

Stellaria media   S21, S23
Stellaria palustris   S24, S27
Stratiotes aloides   **A4**, A11
Suaeda maritima   S4, S21
Subularia aquatica   A13, A22, A23
Succisa pratensis   S24
Symphytum officinale   S7, *S24*, S25

Thalictrum flavum   *S24*
Thelypteris palustris   *S24*, S25
Trifolium repens   S19
Triglochin maritima   S4, S19, S20, S21
Typha angustifolia   S4, S6, **S13**, S14, S20, S24, S25, S28
Typha latifolia   S3, S4, S6, S7, S8, **S12**, S13, S14, S15, S17, S18, S19, S20, S22, S23, S24, S26, S27

Urtica dioica   S3, S4, S5, S6, S7, S12, S14, S15, S23, S24, S25, **S26**, S28
Utricularia intermedia   A13, S2, S4
Utricularia minor   A13, A24, S2, S4, S8, S19

Valeriana dioica   S24, S25
Valeriana officinalis   S6, S7, S11, *S24*, S25, S26, S27
Veronica anagallis-aquatica   S27
Veronica beccabunga   A8, A17, S7, S14, *S23*, S27
Veronica catenata   S23
Veronica scutellata   S11, S23, S27
Vicia cracca   S18, *S24*, S25, S26
Vicia sativa nigra   S18
Viola palustris   S1, S3, S21

Wolffia arrhiza   A1, A3, S14

Zannichellia palustris   A1, A3, A6, A8, A11, A12, A13, A21

fen woodland communities and contact communities. *Journal of Ecology*, **68**, 761–88.

Wheeler, B.D. (1980*d*). Wetland. In *The Magnesian Limestone of Durham County*, ed. T.C. Dunn, pp. 53–60. Durham: Durham County Conservation Trust.

Wheeler, B.D. (1983). A manuscript copy of a chapter 'British Fens: A review' now published in *European Mires*, ed. P.D. Moore, pp. 237–81. Academic Press (1984).

Wheeler, B.D. & Giller, K.E. (1982*a*). Species richness of herbaceous fen vegetation in Broadland, Norfolk, in relation to the quantity of above-ground material. *Journal of Ecology*, **70**, 179–200.

Wheeler, B.D. & Giller, K.E. (1982*b*). Status of aquatic macrophytes in an undrained area of fen in the Norfolk Broads, England. *Aquatic Botany*, **12**, 277–96.

White, J. & Doyle, G. (1982). The vegetation of Ireland: a catalogue raisonné. *Royal Dublin Society Journal of Life Sciences*, **3**, 289–368.

Wiegleb, G. (1978). Untersuchungen über den Zusammenhang zwischen hydrochemischen Umweltfaktoren und Makrophytenvegetation in stehenden Gewässem. *Archiv für Hydrobiologie*, **83**, 443–84.

Wigginton, M.J. & Graham, G.G. (1981). *Guide to the Identification of some Difficult Plant Groups*. Banbury: Nature Conservancy Council, England Field Unit.

Willis, A.J. (1967). The genus *Vulpia* in Britain. *Proceedings of the Botanical Society of the British Isles*, **6**, 386–8.

Willis, A.J. & Davies, E.W. (1960). *Juncus subulatus* Forsk. in the British Isles. *Watsonia*, **4**, 211–17.

Willis, A.J. & Jefferies, R.L. (1959). The plant ecology of the Gordano Valley. *Proceedings of the British Naturalists Society*, **31**, 297–304.

Wolff, P. (1980). Die Hydrilleae (Hydrocharitaceae) in Europa. *Göttinger Floristische Rundbriefe*, **14**, 33–56.

Wolseley, P.A., Palmer, M.A. & Williams, R. (1984). *The Aquatic Flora of the Somerset Levels and Moors*. Peterborough: Nature Conservancy Council.

Woodhead, N. (1951*a*). Biological Flora of the British Isles: *Lobelia dortmanna* L. *Journal of Ecology*, **39**, 458–64.

Woodhead, N. (1951*b*). Biological Flora of the British Isles: *Subularia aquatica* L. *Journal of Ecology*, **39**, 465–9.

Wotek, J. (1974). A preliminary investigation on interactions (competition, allelopathy) between some species of *Lemna*, *Spirodela* and *Wolffia*. *Bericht über das Geobotanische Forschungsinstitut Rübel in Zürich*.

Yapp, R.H. (1908). Sketches of vegetation at home and abroad. IV. Wicken Fen. *New Phytologist*, **7**, 61–81.

Zöbrist, L. (1935). Pflanzensoziologische und bodenkundliche Untersuchung des Schoenetum nigricantis im nordostschweizerischen Mittelande. *Beiträge zur geobotanischen Landesaufnahme der Schweiz*, **18**, 1–144.